An Elementary Survey of
CELESTIAL MECHANICS

An Elementary Survey of
CELESTIAL MECHANICS

Y. RYABOV

Translated by
G. YANKOVSKY

DOVER PUBLICATIONS, INC.
Mineola, New York

Copyright

Copyright © 1961 by Dover Publications, Inc.
All rights reserved.

Bibliographical Note

This Dover edition, first published in 1961 and republished in 2006, is an unabridged republication of *Celestial Mechanics* (a translation of Движения небесных тел) by IUrii Aleksandrovich Riabov, originally published by Foreign Languages Publishing House, Moscow, in 1959.

International Standard Book Number: 0-486-45014-7

Manufactured in the United States of America
Dover Publications, Inc., 31 East 2nd Street, Mineola, N.Y. 11501

CONTENTS

INTRODUCTION

The sky above our heads is populated with stars, planets, the moon, the sun and other luminaries; they all move over the celestial sphere day after day, year in year out, from one century into the next, describing intricate interlacing paths in the heavens. The complex movements of celestial bodies are but the apparent reflection of their yet more complicated actual motions through the endless realms of the universe. And this earth of ours is likewise a celestial object plying its way through space.

How do the heavenly bodies move, and how are these movements related to one another? What forces of nature govern such motions?

Today we can give rather complete answers to these queries. We now know that the earth and planets move in space about the sun forming what is known as the *solar system*, that the sun itself is a member of a huge system of stars called the Galaxy, and, together with other stars, is in motion in space round the centre of this system. We now know that the motions of the earth, planets and sun and stars are governed mainly by forces of mutual attraction. The law of this interaction—the law of universal gravitation—was discovered in the seventeenth century by the great English scientist Isaac Newton.

Utilizing the law of universal gravitation, celestial mechanics—the science that treats of the motions of celestial bodies—has achieved remarkable results. We now are able to draw up exact "schedules" of the motions of heavenly bodies with indications of where, in what part of the sky, any given body will be at any given instant of time. And true enough, celestial objects arrive at the required "stations" in the sky at exactly the prescribed time in strict accord with our time-table, probably more exactly than do the trains of a terrestrial railway line. The astral time-table is drawn up not for a day or a year, but for tens and hundreds of years in advance. In some cases we can even draw the picture of heavenly movements that our distant ancestors viewed millennia ago, and we can look into a future so far off that generations will come and go before this picture comes to pass.

Naturally, celestial mechanics was not created overnight. Long, centuries-long was the search for the truth, and on the way were errors, delusions—and the struggle for this truth was oft-times fierce. Even today we are far from a complete knowledge of the motions of celestial bodies. And not for all objects in the sky can we draw up a sufficiently accurate "time-table" of their movements. Occasionally our schedule goes awry—some bodies move too fast, some fall behind. And we are not always in a position to say exactly and definitely how a certain body moved ages ago or will move in the distant future.

The aim of this book is to tell the reader how the law of universal gravitation was discovered and how, on the basis of this law, the motions of celestial bodies are studied. We will learn that the movements of the most distant stars and the falling of bodies to the ground and the flight of man-made satellites of the earth, and of cosmic rockets too, are all subject to the same law of gravitation.

Towards the end we shall deal in brief with the nature of gravitation.

1. ANCIENT CONCEPTIONS CONCERNING THE MOTIONS OF THE SUN, MOON, PLANETS AND STARS

The movements of heavenly bodies began to be studied long, long ago. During many centuries before our time ancient peoples observed the positions and movements of celestial objects in the sky and strived to notice regularities in these motions. The stimulus that guided them was primarily practical necessity. It was only by observing the heavenly bodies that one could find his bearings in the desert or at sea; or could measure time and predict the season of the year. The demands of trade and agriculture and the migrations of nomads all required observations of the stars. Thus it was that urgent practical demands engendered astronomy, the science of celestial objects.

The apparent motions of heavenly bodies across the sky were well known even in very ancient times.

Watching the starry sky at night one gets the impression of a dome-shaped celestial sphere in rotation about the earth making a single circuit every 24 hours. This diurnal rotation of the firmament is repeated regularly, from day to day, without any apparent change.

The stars in the sky appear firmly fixed in place with respect to each other. Hence the name *fixed stars*. Ages ago astronomers had already compiled permanent maps of different constellations and of the entire stellar sky.

All heavenly bodies without exception participate in the diurnal rotation about the earth. But if a given body retains its relative position among the stars it is called a fixed star. In contrast, when we speak of an object moving across the sky we have in mind not its diurnal motion but its movement with respect to the fixed stars.

Antiquity knew of seven bodies that moved among the stars. They were called *planets* (which in Greek meant "wanderers"). Two points should be noted in this respect. First, these included five bodies that were far brighter that the stars, and the names given them by the ancient Romans—Mercury, Venus, Mars, Jupiter, and Saturn—remain to this day. Secondly, in ancient times, the planets included the sun and moon too for they likewise were in motion among the stars.

The moon's motion is easiest to notice because this body moves faster than the other objects. The moon moves from west to east and makes a complete circuit across the sky in only a little over 27 days (which amounts to a speed of 12-13 degrees per 24 hours or $0°.5$ an hour). One needs only two nights to notice that the moon has changed its position relative to the stars. Observations show that the moon's movement across the sky is not uniform, on some sections it moves faster, on others slower.

The sun's movement among the stars cannot be registered directly since the latter are not visible in the day-time, but such movement can be detected from observations of the stars. Note some star in the western part of the sky soon after sunset and then try to locate it a few days later at the same hour. You will find that it has moved down closer to the sun. In another few days it will disappear altogether below the horizon, and its place will be taken by another star lying to the west. This new star will gradually approach the sun and continue as the first, etc. This shows that the sun changes its position relative to the stars. Way back in ancient times daily observations were made of the positions of the stars on the celestial sphere. These observations permitted of a rather accurate study of the sun's path across the sky. It was found that the sun, and the moon too, moves among the stars from west to east, describing a complete circuit in roughly 365 and a quarter days, that is, in one year. The annual path of the sun among the stars remains constant from year to year and is known as the *ecliptic*. In a single day the sun moves eastward along the ecliptic approximately one degree (or 360° per year). The sun's movement, again like that of the moon, is not uniform throughout the year. In winter it moves faster than in summer. For example, between June 1 and June 30 it covers 27°.5, while between December 1 and 30 it does 29°.5.

The picture of the apparent motions of the planets is more complicated. A common feature that the planets have is that they always move close to the ecliptic. The planets fall into two groups: the inferior planets and the superior planets.*

In the first group belong Mercury and Venus, in the second, all the others. Characteristic of Mercury and Venus is that they oscillate about the sun as about a certain mean position. At first the planet moves among the stars more rapidly than the sun and passes it; then, having moved to the farthest point east of the sun, the planet begins to move more slowly than the sun, which overtakes it. After lagging behind a certain distance to the west, the planet again begins moving faster than the sun, and the cycle is repeated. For Venus, the maximum solar distance is about 40°, while for Mercury it is an average of 23° (varying from

---

* These designations have come down to us from ancient times when it was believed that the sun and all the planets moved about the centre of the Universe—the earth. It was thought that Mercury and Venus were closer (lower) to the earth than the sun, and that the other planets were farther away (higher).

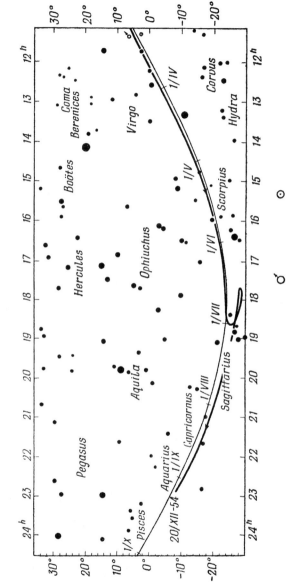

Fig. 1. The apparent paths of Mars ($\male$) from October 10, 1953, to December 20, 1954, and the sun ($\odot$) between March and October

18 to 28°). For this reason, these planets are visible only in the morning in the east shortly before sunrise and in the evening in the west just a little after sunset.

The period of apparent oscillations of Mercury about the sun is 116 days, for Venus it comes out to one year and 217 days.

The superior planets describe characteristic loops in the sky.

The overall motion of the planets among the stars is from west to east like the sun and moon (this is called *direct motion*). However, there are times when the rate of apparent motion of a planet diminishes, the planet comes to a halt, and then begins to move from east to west (*retrograde motion*). Each planet has its own period of retrograde motion with a subsequent reversal of direction to direct motion from west to east. The result is a peculiar type of loop that the planet describes in the sky (Fig. 1). Mars describes a loop every 780 days, Jupiter every 399 days, Saturn every 378 days.

These are the movements of the stars and planets on the sky as viewed from the earth. But what are the real movements of the stars, sun, moon and planets in space?

The first attempts to explain the observed motions of celestial bodies, to build a theory of their motions that could predict the location of a given body in the sky at a given instant were made by learned men in ancient Greece. Their starting point was a stationary earth with the sun, moon, planets and stars revolving about it.

The ancient Greeks contrasted "terrestrial" and "celestial" phenomena and believed that the laws of the "heavens" should differ utterly from those of the "earth." The ever recurring motions of celestial objects appeared to them a paragon of perfection, and since they regarded uniform motion along a circle the most perfect type of motion, it seemed to them an immutable fact that the moon, sun and planets should move in circles at a uniform rate. However, the apparent motions of these bodies hardly resemble uniform circular movements. The sun and moon have rates of motion that are not uniform, while the planets even describe intricate loops. Ancient Greek astronomy was thus confronted with the following problem: to give an explanation of the apparent movements of the planets based on uniform circular motion.

The most refined theory of the motions of heavenly bodies of that time was developed in the second century A.D. by the ancient Greek astronomer Claudius Ptolemy. He supposed the sun, moon and planets to be in uniform motion in circles that were called *epicycles*. In turn, the centre of each epicycle was in uniform motion along a larger circle, known as the *deferent*, with the motionless earth at its centre. In this scheme, the rate of motion of the planet V_1 along the epicycle is greater than the rate of motion V_0 of the centre of the epicycle along the deferent. During the interval when the planet P and the centre of the epicycle O are moving in a single direction, the observer on earth E sees direct motion. But if the planet is in motion between the centre of the epicycle and the earth, the motions of the planet and the centre of the epicycle are "deducted" one from the other, and, seeing that $V_1 > V_0$, the planet will be in retrograde motion when viewed from the earth.

By selecting for the sun and the moon and each planet the ratios of deferent and epicycle radii, the orbital periods along epicycle and deferent, and the mutual inclinations of the planes of the deferent and epicycle, Ptolemy was able not only to explain the nonuniform apparent motion of these bodies across the sky and the retrograde movements of the planets, but even to compute rather accurately the paths of the planets, sun and moon across the heavens.

The Ptolemaic system and in general the teaching that made the earth a stationary body was unrivaled for fourteen centuries, from the second to the middle of sixteenth century. True, even before Ptolemy, the ancient Greek scholars Philolaus (fifth century B.C.) and Aristarchus of Samos (third century B.C.) and some others propounded the view that the earth is in motion in space and that, in addition, it rotates on its axis; on their view, the observed diurnal rotation of the celestial sphere and of all the stars is simply a reflection of the actual rotation of the earth. But these thinkers were unable to offer methods of precise calculations to pre-

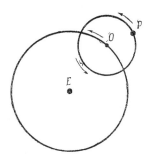

Fig. 2. An epicycle. The velocity of V_1 is greater than that of V_0

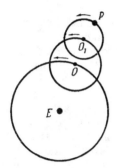

Fig. 3. A compounded epicycle

dict the planetary positions in the heavens and thus satisfy the practical demands of astronomy, and therefore their brilliant guesses did not become generally accepted.

This long period of domination of the Ptolemaic system was of course due not only to the low level of science at that time. The point is that an earth-centred universe was ideally suited to religion. Since man is the "crowning achievement of God" it is natural that man's abode—the earth—should be located at the centre. Thus it was that Ptolemy's cosmology served religion. Quite natural then that the pagan priests accused Aristarchus of Samos of godlessness for his teaching of the earth's motion, just as, eighteen centuries later, Christian churchmen fought furiously against the Copernican system.

But finally the Ptolemaic system came to an impasse, for as new observational data accumulated, more discrepancies were detected between observed planetary positions and those predicted by Ptolemy's theory. To eliminate these divergences, it was assumed that the motion of each planet is to be described not by a single epicycle but by a whole system of epicycles (Fig.3). Each newly discovered irregularity in the motion of a planet was eliminated by the addition of more and more epicycles. Yet disparities between theory and observation remained.

This extremely unwieldy and obviously artificially compounded system of epicycles, on the one hand, and the lack of complete agreement between theory and observation, on the other, finally resulted in doubts cropping up as to its validity. The time came when astronomy ceased to "refine" Ptolemy's system by adding more and more epicycles, and rejected it outright. This was the sixteenth century, when the great Polish astronomer Nicolaus Copernicus (1473-1543) created the first complete theory of the motion of the earth and planets about the sun. This is actually the starting point of our knowledge about the true motions of celestial bodies in space.

2. THE GEOMETRY OF PLANETARY MOTIONS FROM COPERNICUS TO KEPLER

Copernicus discarded the dogmatic assertion of a stationary earth that had dominated the minds of men for ages. According to his theory, the earth is in no way different from the planets in its motion in space about the sun and in its rotation round a certain imaginary line which we call the earth's axis. The diurnal movements of the stars and all other heavenly bodies on the celestial sphere were correctly explained by Copernicus as the result not of their actual motions but of the rotation of the earth. The earth rotates, completing a circuit in 24 hours, while to a man on the earth's surface who does not feel any motion it appears that the entire celestial sphere with the sun, stars, and planets attached is in rotation.

The annual path of the sun is, according to Copernicus' system, only the apparent motion produced by the earth's movement in space about the sun. The earth circles the sun, and earth dwellers see the sun on the background of different stars that are at far greater distances than the sun. This is why it seems to us that the sun moves among the stars.

Copernicus demonstrated that the principal peculiarities in the apparent planetary motions can be explained by the fact that the planets, including the earth as one, move about the sun in one and the same direction at different distances from it and make complete circuits in definite times.

Reasoning from observational facts, Copernicus first came to the conclusion that all the planets and the earth move round the sun in approximately the same plane. This explained why the paths of the planets, as seen from the earth, lie near the ecliptic. Inasmuch as Mercury and Venus seem to oscillate about the sun, their paths in space, or, in astronomical parlance, their, *orbits*, lie closer to the sun than that of the earth; and Venus is farther from the sun than Mercury because its apparent deviations from the sun are greater.

The other planets revolve about the sun at greater distances than the earth. Closest to the earth is Mars (this is evident from the fact that it moves fastest among the stars), followed by Jupiter and then Saturn.

As regards the shapes of the planetary orbits and the

type of motion of the planets, Copernicus was of the opinion that all planets exhibited a nearly uniform motion in circles but that additional oscillations were superimposed on these uniform circular movements. More precisely, Copernicus thought that it was not the planets themselves that moved uniformly in circles but the centres of epicycles or

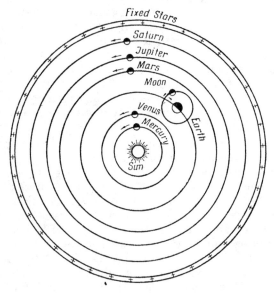

Fig. 4. Copernicus' World

systems of epicycles along which the planets proper moved. Motion along epicycles characterized the deviations from uniform circular motions about the sun.

Copernicus was the first astronomer to produce a correct plan of the solar system. He determined the relative distances of the planets from the sun (in terms of the earth-sun distance) and also their periods about the sun. Here are his calculations.

Consider, for example, the planet Mercury, which is closer to the sun than the earth. In Fig. 5, the inner circle is the orbit of Mercury, the outer circle, that of the earth; arrows indicate the direction of motion. From the figure it is clear that when viewed from the earth Mercury should always be close to the sun oscillating about the latter. E_1 and M_1 denote the earth and Mercury when the latter is farthest from the sun to the west. The angular distance* between the sun and Mercury is then an average of about 23° (Copernicus explained the oscillations in angular distance between Mercury and the sun—from 18° to 28°—with the aid of epicycles). Since the triangle

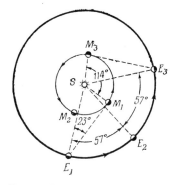

Fig. 5. Determining Mercury's distance from the sun and its orbital period about the sun

$SE_1 M_1$ is a right-angle triangle, by trigonometry we obtain:

$$\frac{SM_1}{SE_1} = \sin 23° \approx 0.39.$$

Thus it turns out that Mercury's mean distance from the sun is less than that of the earth by a factor of roughly 2.6.

Earlier it was pointed out that the period of Mercury's apparent oscillations about the sun is approximately 116 days. This means that in just about 58 days Mercury will again be seen at maximum elongation from the sun, but this time to the east. However, the earth, and hence also Mercury, will occupy different positions in their orbits. Let us designate these positions by E_2 and M_2. The length of the arc E_1E_2 may easily be found since it is known that the earth makes one complete circuit about the sun in 365.25 days. In 58 days the earth covers roughly 0.159 of its orbit, an arc of 57°. In another 58 days Mercury will again be seen at maximum elongation to the west. We denote the positions occupied by Mercury and the earth at this time by M_3 and E_3. Thus, in 116 days the earth will describe an arc E_1E_3, that is, 57°+57°=114°. During this time Mercury does more than one circuit round the sun so that the

---

* Angular distance is the angle between two lines pointing from the observer's eye to the two celestial bodies whose separation is to be measured.

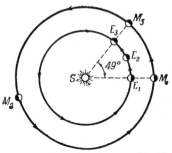

Fig. 6. Determining the Martian period of revolution

period of apparent oscillations of Mercury about the sun does not coincide with its orbital period. But the latter can easily be computed.

Indeed, from Fig. 5 it may be seen that the leg SE_1 of triangle SE_1M_1 has moved 114° to the position SE_3, and therefore the leg SM_1, at position SM_3, has moved 114°. Hence $\angle M_1SM_3 = 114°$. Thus, in 116 days Mercury describes a complete circle plus 114°, or an arc of 474°. Mercury's orbital period, that is, the time it takes to complete one circuit (360°) may be found from the ratio.

$$\frac{T}{116} = \frac{360°}{474°}$$

Whence $$T = \frac{360 \times 116}{474} \approx 88 \text{ days.}$$

In exactly the same way we can calculate the distance of Venus from the sun and also the planet's orbital period. Venus has an orbital period of 225 days and a mean solar distance 0.72 that of the earth-sun distance.

A different method can be used to determine the orbital periods and solar distances of planets farther away from the sun than the earth.

Take Mars, for instance. In Fig. 6 the inner circle is the earth's orbit, while the outer one is that of Mars. We denote by E_1 and M_1 the positions of the earth and Mars at a time when Mars and the sun are on a straight line at opposite sides of the earth (this configuration is known as *opposition*). Observations show that oppositions of Mars occur every 780 days. Three hundred and ninety days after opposition, the earth will be at E_2 and Mars will be on one line with the sun and on the same side from the earth (this position is called *conjunction* of the planet and the sun). In another 390 days Mars will again be in opposition to the sun, and the earth will be at E_3 and Mars at M_3. In 780 days the earth sweeps out two complete circuits about the sun plus the arc E_1E_3, which amounts to about 49°, in other words, a total of 769°, while Mars, as may be seen from the figure, completes one circuit plus an arc of 49°, making a total of 409°. By means of ratios we obtain the following orbital period for Mars:

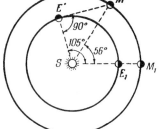

$$T = \frac{360 \times 780}{409} \approx 687 \quad \text{days.}$$

Fig. 7. Determining the distance of Mars from the sun

The method is the same for determining the orbital periods of Jupiter and Saturn, which come out to 12 years and 29.5 years respectively.

Now let us determine the distance of Mars from the sun. Beginning with opposition we measure the angle between the sun and Mars as seen from the earth (Fig. 7). At opposition this angle is 180°. It begins to diminish and there comes a time when it reaches 90° (E' and M'). Observations show that this occurs roughly 106 days following opposition. During this time the earth has moved through an arc of about 105°, while Mars has completed approximately $\frac{360}{687} \times 106 \approx$ $\approx 56°$; the angle $E'SM'$ will be $105° - 56° = 49°$. From the triangle $E'SM'$ we obtain

$$\frac{SM'}{SE'} = \frac{1}{\cos 49°} \approx 1.5.$$

Thus, Mars is one and a half times more distant from the sun than the earth.

In similar fashion we can obtain the relative distances of Jupiter and Saturn from the sun. Calculations show that Jupiter moves at a distance from the sun five times that of the earth's solar distance, while that of Saturn is 9.5 times farther away.

Copernicus demonstrated that the loop-like apparent paths of the planets could be explained by the fact that we observe the planets from the earth which itself is in motion about the sun. By way of illustration, let us see how Mars moves beginning from the point of opposition (E_1M_1 in Fig. 8). Within a certain interval of time following opposition, the earth will reach E_2, while Mars during the same interval will describe in its orbit a smaller arc M_1M_2 (since its orbital period is longer). At opposition Mars was in a direction E_1M_1, some time after opposition it will be seen in the direction E_2M_2, which moved to the *west* with respect to the direction E_1M_1. The earth appears to overtake Mars in its motion about the sun, creating the impression that Mars is moving among the stars

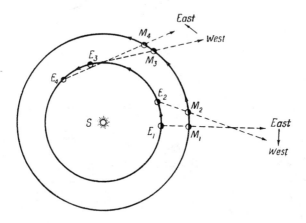

Fig. 8. The apparent motions of the planets explained

westward from its original position, though in reality Mars continues to move in its orbit in the same direction as the earth.

Let us see how the earth-Mars line will have changed when the two planets reach E_3 and M_3, that is, when Mars and the sun form a right angle as seen from the earth (Fig. 8). Some time later the earth reaches E_4 and Mars M_4. E_4M_4 now points *eastward* relative to E_3M_3, which means that Mars will be moving among the stars from west to east. Thus, during the time that the earth described in its orbit the arc E_1E_4, Mars (speaking of apparent motion) first moved from east to west among the stars (*retrograde motion*) and then from west to east (*direct motion*). At a certain point in between M_1 and M_4 Mars will come to a "stop" (E_s and M_s in Fig. 9) and reverse its path among the stars. Observations show that Mars comes to a halt roughly 35 days following opposition. Mars then reverses its direction to direct motion. But approximately 35 days prior to the next opposition, when the earth and Mars are at E'_s and M'_s (the mirror images of E_s and M_s)—see Fig. 9—Mars again reverses its direction to retrograde motion. Thus, retrograde motion continues 35 days prior to opposition and 35 days following opposition, making a total of 70 days.

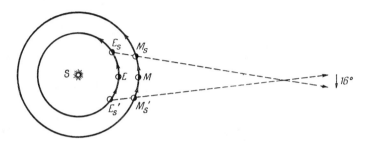

Fig. 9. The retrograde motion of a planet

The arc of retrograde motion which Mars describes during this time amounts to roughly 16°.

In this way Copernicus explained the retrograde motions of Mars, Jupiter and Saturn. But in explaining the other planetary deviations from uniform circular motion Copernicus retained the system of epicycles of the ancient astronomers. Particularly involved was Copernicus' theory of the moon, whose motion was found, by the astronomers of antiquity, to contain many irregularities.

Nevertheless, the Copernican system was a new stage in the development of astronomy. Copernicus was the first to put the sun, earth, and planets in their places. He created, on the whole, a correct picture, an orderly system of the earth's motion and that of the planets. He succeeded in defining the most important characteristics of planetary motion: the orbital periods and solar distances of the planets.

But the significance of Copernicanism extends far beyond the pale of astronomy proper. Copernicus was the first to demand that astronomical schemes should reflect reality, "the true nature of things," and not the teachings of religious systems. Thus it was that Copernicus dealt a severe blow at the world view of religion that was based on the writings of the church fathers, and his name became the banner of progressive science.

No wonder that Copernicus' theory was so hostilely received by the church—it reduced the earth to the status of an ordinary planet and placed the sun at the centre of the solar system. The earth was no longer the centre of the world as the Bible had taught. In 1616 the Catholic church pronounced the Copernican system heretical and banned it. Even before this time there had been active opposition to the new theory by prominent representatives of protestantism.

All of this, of course, greatly hindered the spread of Copernicus' theory. One should likewise bear in mind that Copernicus had no direct proof of the earth's motion. The only thing that he could say was that his theory gave a simpler and more natural explanation to the apparent motions of the planets than did the Ptolemaic system. Moreover, the Copernican system was not yet able to predict planetary position with sufficient accuracy.

Nevertheless, Copernican views gradually came to be accepted. An exceedingly important part in the development and victory of the new scheme of the solar system was played by two remarkable men, the Italian Galileo Galilei (1564-1642) and the German Johannes Kepler (1571-1630).

In 1610, for the first time in the history of astronomy, Galileo directed a telescope to the heavens, thus opening up unimaginable vistas for astronomical exploration. His very first observations resulted in a series of remarkable discoveries. First of all, he found that Jupiter had four small stars circling it. The shortest orbital period of these stars was 42 hours, the longest, 17 days. Galileo christened them the "Medicean planets" in honour of Cosimo de' Medici, Grand Duke of Tuscany. It was not long before Kepler gave them the name of *satellites*. This name has remained to the present day for all celestial bodies that revolve round their primaries as the planets do about the sun.

The "moons" of Jupiter that Galileo discovered reproduced in miniature the sun's system of planets. The new satellites proved the old dogma—that only the stable earth could be the centre of motion—to be wrong. Jupiter's

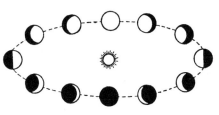

Fig. 10. The phases of Venus

orbiting satellites graphically destroyed the "objection" to Copernicus' system that the moon could not revolve about a moving earth without lagging behind it.

Galileo also found that Venus does not always appear as a full disk, but, like the moon, changes its appearance— first crescent, then full, and at times it is not seen at all. This was proof that, like the moon, Venus is a dark body that receives its light from the sun. The changes in Venusian phases occurred exactly as might be predicted if it were in orbital motion about the sun (Fig. 10).

Buttressed by these and a whole series of other discoveries, Galileo was highly successful in spreading Copernican astronomy both from the university rostrum and in a number of masterly written compositions. For this he was summoned before the tribunal of the Inquisition and in 1633 forced to recant publicly his "fallacies."

The next step in the study of the planetary motions proper was made by the pre-eminent work of Kepler, Galileo's contemporary.

By about 1600 the tables of planetary motions compiled on the basis Copernicus' theory were predicting planetary positions with errors up to 4° and 5°. Such gross errors showed that the Copernican system was anything but perfect. Kepler himself was an ardent follower of Copernicanism and least of all thought of abandoning its basic principles. The only thing he questioned was the correctness of the Copernican-Ptolemaic system of epicycles. Kepler had at his disposal extensive and precise observations of the planet Mars obtained by the Danish astronomer Tycho Brahe (1546-1601)—most proficient observer of that period. Using Mars as an example, Kepler set out to make a detailed study of the nature of planetary motions.

Since the apparent motion of Mars is due both to the motion of Mars itself and of the earth, Kepler decided first to define precisely the earth's orbit. To do this he applied an ingenious technique that permitted him first to study the irregularities in the earth's orbit. Its underlying principle consists in the following.

Let us suppose that Mars is observed at intervals of time following opposition equal to one, two and several orbital periods of the planet. Mars will then always be in one and the same position in its orbit, while the earth will occupy different positions (see Fig. 11, M refers to Mars and E_0, E_1, E_2, etc., refer to the earth).

If the earth is at E_0 at the first observation, then in 687 days (the Martian orbital period) it will not have time to complete two full circuits and will be at E_1. After the next 687 days it will be at E_2, and so on.

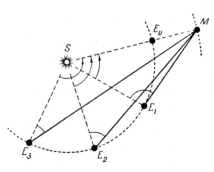

Fig. 11. Determining the earth's orbit

First of all, note that it is possible to determine the angles between different positions of the sun as seen from the earth by comparing the apparent solar positions on the ecliptic at different instants. Indeed, the apparent distance covered by the sun along the ecliptic in degrees during a certain time interval will be equal to the angle swept out by the sun-earth line during this time. The motion of the sun along the ecliptic had been studied by the ancient Greeks and by Tycho Brahe. So Kepler had at his disposal a rather precise set of tables of solar motion with indications of where the sun should be on the ecliptic each day.

Thus, the angles E_0SE_1, E_0SE_2, etc., were known to Kepler. From observations one could find the angles SE_1M, SE_2M, etc., between the sun and Mars.* Then, from tri-

---

* These angles are not usually determined directly, since Mars and the sun are rarely visible in the sky at one time, but by measuring the positions of these bodies relative to the stars, the angular distances between which are known.

angles SME_1, SME_2, SME_3, etc., in
which one side (SM) is always constant
and two angles are known, it is possible by trigonometry to determine
the distances SE_1, SE_2, etc., in fractions of the distance of SM and en-
ter, on the drawing, points that correspond to the positions of the earth
at different instants. It is then possible to use these points to draw
a curve that depicts the path of the

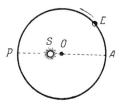

Fig. 12. Kepler's concep-
tion of the earth's orbit

earth about the sun.

It was found that this curve is a circle with the sun not
exactly in the centre but slightly displaced from it. (Fig.
12.) The distance OS between the centre of the circle and
the position of the sun was, according to Kepler's figures,
roughly 1/59 the radius of the circle. This is a very small
value. In a circle of radius 10 cm. this displacement would
amount to only 1.5 mm.

Kepler also noticed that the earth does not move uniformly in its orbit. When the earth is closest to the sun near
P in Fig. 12—this point in the orbit is called the *perihelion*—it moves faster than when it moves away from the
sun and approaches A (this point of the orbit most distant
from the sun is known as the *aphelion*). Taking into account
this irregularity, Kepler compiled a detailed table of the
earth's orbital motions with the positions of the earth
given for nearly every day of the year.

Having completed this work, Kepler began a revision of
the Martian orbit. From Tycho Brahe's observations he
selected those that gave
the positions of the planet
every one or several of
its orbital periods.

Fig. 13 shows the po-
sitions of the earth and
Mars at opposition (E_0, M)
and after a single Martian
orbital period (E_1M). Kepler could determine the
angle φ and the distance SE_1
from the tables of terres-
trial motion that he had

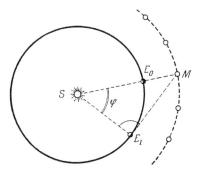

Fig. 13. Determining the orbit of Mars

compiled.. Angle SME_1 was determined from observations. In this way, from triangle SME_1 it was possible to find SM, the distance of Mars from the sun.

In a similar manner Kepler computed the distance of Mars from the sun at different points in the orbit and attempted to find a curve that would pass through all these points. However, after a long time and arduous labour

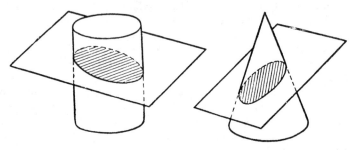

Fig. 14. When a plane cuts a cylinder or a cone it produces an ellipse

Kepler concluded that the Martian orbit could not be a circle and that the motion of this planet could not be represented by a combination of circular motions. Thus it was that the view which had predominated for centuries— celestial motions could only be circular motions—was disproved. Kepler then tried to draw an elongated curve through these points and found—again after numerous trials—that the Martian orbit could very well be represented by an ellipse, the simplest type of oval curve resembling an elongated circle with the sun at one of the foci.*

We know that the ellipse is a curve, for any point of which the sum of its distances from two given points called the *foci* of the ellipse is a constant. An ellipse may be obtained by an inclined plane cutting a cone or cylinder. (Fig. 14).

In Fig. 15, F_1 and F_2 are foci of the ellipse and O is the centre. AD is the major axis and BE the minor axis

---

* This is a simplified version of Kepler's discovery. The actual reasoning was far more involved.

24

of the ellipse. $AO=OD$ and $BO=OE$ are the semi-major and semi-minor axes of the ellipse respectively. AD is known as the *line of apsides*. The ratio $e=\dfrac{OF_1}{OA}=\dfrac{OF_2}{OA}$ is called the eccentricity of the ellipse. The greater the eccentricity the more displaced are the foci from the centre and the greater is the difference between the semi-major and

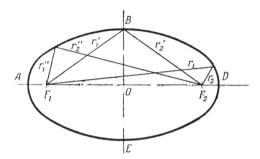

Fig. 15. An ellipse. The sums $r_1+r_2=r'_1+r'_2 = r''_1+r''_2$ are constant and equal to the major axis of the ellipse

semi-minor axes. The distance of the centre of the ellipse from the foci is calculated from the equation $OF_1=OF_2==e\times OA$. The semi-minor axis BO is related to the semi-major axis AO and the eccentricity e by the equation:

$$BO = AO \sqrt{1 - e^2}.$$

The more BO differs from AO the more elongated is the ellipse and the more it differs from a circle. When the eccentricity is small the semi-major and semi-minor axes are nearly the same, and the ellipse differs but slightly from a circle.

From the equation relating AO, BO and e it may be seen that if, for example, the eccentricity is 0.1 and $AO=10$ cm., that is, the foci are displaced from the centre 1 cm., the difference between the semi-major axis and the semi-minor axis is only 0.05 cm.

Kepler computed the eccentricity of the elliptical orbit of Mars at about 1/11. The semi-major axis of the Martian orbit was 1.52 times that of the radius of the circle in

which the earth revolved. Fig. 16 illustrates an elliptical
orbit with this eccentricity. Judging from the drawing it
is very difficult to distinguish the ellipse from a circle,
but the displacement of the sun with respect to the centre
of the ellipse is readily apparent.

Observations showed that Mars appears to oscillate about
the ecliptic. Earlier this was explained by means of ad-

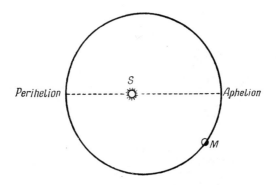

Fig. 16. The Martian orbit after Kepler

ditional planetary motions in epicycles. And Copernicus
adhered to it too. Analyzing the observations of Tycho
Brahe, Kepler found that all of Mars' deviations from the
ecliptic could be explained by the fact that Mars always
moves in a single plane which is inclined to the earth's
plane. This angle is known as the *inclination* of the planetary
orbit. For Mars it is roughly 2°.

Summarizing, Kepler established the fact that Mars
moves in an elliptical orbit in a plane inclined 2° to the
plane of the earth's orbit, with the sun in a focus of this
orbit.

This was the first in a series of remarkable regularities
that Kepler discovered in the motion of Mars.

Kepler then made a study of the peculiarities of
the planet's orbital motion. He found that near perihelion
Mars covers in two months a path of 37°.0 as seen from
the sun, while during the same period of time near aphelion
Mars sweeps out an arc of only 25°.8. Thus, the farther Mars
is from the sun the slower it moves in its orbit: near peri-

helion the speed is greatest, near aphelion, the least. Kepler tried a large number of hypotheses and at last hit upon a remarkable regularity (Fig. 17): Mars moves in its orbit in such a way that if we take the sections of its path M_1M_2, M_3M_4, etc., that Mars sweeps out during one and the same time in different parts

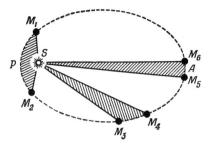

Fig. 17. The "Law of Areas": if the planet sweeps out the arcs M_1M_2, M_3 M_4, M_5M_6 in equal intervals of time, the areas of the shaded sectors are equal

of the orbit, and connect the ends of the arcs with the sun, the areas of these sectors (SM_1M_2, SM_3M_4, etc.) will be equal.* This regularity may be expressed otherwise: the areas swept over by the *radius vector* (line joining the centre of the sun with the centre of the planet) of the planet are proportional to the times.

This was the second remarkable regularity in the motion of Mars—the *Law of Areas*. Thus, for Mars two important laws were found that are now known as the *First* and *Second* Keplerian Laws.

1. The planet moves in an ellipse with the sun at one focus.

2. The straight line connecting the planet with the sun (the radius vector of the planet) passes over equal areas in equal intervals of time.

Although Kepler established these two laws only with respect to Mars he was convinced that they hold for all planets. For one thing, Kepler could straightway verify the validity of his laws as regards the earth. Indeed, according to Kepler's data the earth moves in a circle with the sun displaced from the centre 1/59 the radius. This, then, should be the eccentricity of the earth's elliptical orbit. With an eccentricity so small the semi-minor axis should differ from the semi-major axis by only one part in 7,000. Of course, in his day Kepler was unable to distin-

---

* To make the idea clear, Fig. 17 depicts a much more elongated ellipse than that of the Martian orbit. The lengths of the arcs M_1M_2, M_3M_4, etc., differ greatly from actuality.

guish such an ellipse from a circle.* But at any rate the data on the earth's motion that Kepler had at hand did not conflict with the First Law. As regards the Second Law, Kepler's tables of the earth's motion revealed that the earth moves faster when closest to the sun and slower when farther away, exactly as follows from the Second Law.

Later, Kepler determined the approximate elliptical orbits of the other planets.

Kepler's First and Second Laws were momentous discoveries for the science of the heavens. But Kepler did not give up the hope of finding a general law for all the planets of the solar system that would embrace the prominent fact that the farther a planet is from the sun the slower it moves. This remarkable ratio Kepler discovered only in 1618; it is now known as Kepler's *Third Law*. By correlating the sizes of the orbits of the planets and their orbital periods about the sun, Kepler discovered that the squares of the periods of revolution of any two planets about the sun are to each other as the cubes of the semi-major axes of their elliptical orbits (or, otherwise stated, as the cubes of their mean distances from the sun).

For example, the mean distance of the earth from the sun is to the mean distance of Mars as 1 : 1.52 (according to Kepler's data) and their periods of revolution about the sun are as 1:1.88. If we cube the first ratio and square the second we obtain nearly equal values: 1/3.53 and 1/3.54.

Correlating the mean distances from the sun of the earth and Jupiter we obtain for the ratio of the cubes of the distances 1/125 (it was known that Jupiter is roughly five times farther from the sun than the earth); for the squares of the periods we find 1/144 (Jupiter has an orbital period of about 12 years). The equality of these ratios is naturally only approximate since the distance of Jupiter from the sun and its period of revolution were rather imperfectly known at that time. But Kepler regarded his law as absolutely exact.

If we take the earth-sun distance and the earth's orbital period to be unity, and utilize present-day data on the

---

* One should bear in mind that in Kepler's day all observations (including those of Tycho Brahe that Kepler utilized) were naked-eye observations.

distances and orbital periods of the planets we obtain the
following table:

Table 1

Planet	Semi-major axis, a	Orbital period, T	a^3	T^2
Mercury	0.387	0.241	0.058	0.058
Venus	0.723	0.615	0.378	0.378
Earth	1.000	1.000	1.000	1.000
Mars	1.524	1.881	3.540	3.538
Jupiter	5.203	11.862	140.8	140.7
Saturn	9.539	29.458	868.0	867.9

From the table it is seen that for any two planets the
ratio of the cubes of their mean distances from the sun
$(a_1/a_2)^3$ almost exactly coincides with the ratio of the squares
of their periods T_1^2/T_2^2. The reasons for the slight deviations
from the Third Law that are apparent from the tabulated
numbers will be explained later on.

On the basis of the laws that he had discovered and
after many years of painstaking computation, Kepler compiled tables that indicated the position of each planet in
the heavens at any instant of time. These tables were is-
sued in 1627 and proved far superior to all the astronomical
tables in use before. This was a clear demonstration of
the correctness of his laws and justified his place in the
history of astronomy as the "law-giver of the heavens."

We must once again stress the fact that Kepler believed
his laws applicable to the movements of all planets, and
also to those of the moon and the four Jovian satellites
then known. Kepler correlated the distances of Jupiter's
satellites from the planet and their orbital periods and
found that his Third Law held. He applied his first two
laws to the motion of the moon, but it turned out that lunar motion is complicated by a large number of irregularities which Kepler was unable to explain.

Kepler's studies brought to a close the first period in
the investigation of planetary motion, which may be called
the *descriptive* or *geometric* period. *How* the planets
moved was known, and detailed and rather accurate tables
of planetary motion had been compiled, but it was still not
clear *why* the planets moved as they did—in accordance

with Kepler's Laws. What makes the planets move about the sun, Jupiter's satellites round their primary, and the moon round the earth?

The sources of the development of the causal theory of planetary motion—the dynamics of the solar system—are intimately bound to the same glorious names: Copernicus, Kepler and Galileo.

3. THE DISCOVERY OF THE LAW OF GRAVITATION

Everyone knows that all bodies fall to the earth because of the force of gravity. All material bodies that are in no way supported will fall to earth. What makes them fall? The ancient Greeks "explained" this simple fact by saying that all heavy bodies must strive "down," to the centre of the universe, which in antiquity was considered to be the centre of the earth. And it was this property that they called gravity.

As long as the earth was considered something exceptional and unique in the world, weight (or gravity) was believed to be a purely terrestrial phenomenon that had nothing to do with things celestial. But the discoveries of Copernicus and his adherents proved that the earth is an ordinary planet moving about the sun together with the other planets and that the earth is a heavenly body very much like other heavenly bodies. In this connection it occurred that the property of gravity was an attribute not only of the earth but of other celestial bodies as well. If material bodies in the vicinity of the earth strive towards its centre, then these same bodies in the vicinity of the moon, planets or sun should also strive to the centres of these celestial objects too. In other words, they should be *attracted* by these objects.

Copernicus and, later, Kepler correctly supposed that celestial bodies possess the property of attracting.* They regarded this property as the tendency of homogeneous bodies to coalesce. In his book *Astronomia nova* Kepler writes:

"Gravity is the mutual inclination between similar bod-

---

* Similar views on gravity had been expressed earlier by Nicholas of Cusa and Leonardo da Vinci.

ies striving to merge, coalesce. . . . No matter where we put the earth, heavy bodies will always strive to move towards it due to their peculiarity. . . . If two stones were removed to any part of the world, near each other but outside the field of force of a third related body, then the two stones, like two magnetic bodies, would come together. . . ."

Thus, due to an extension of the property of gravity to other celestial bodies, the question now posed was that of the *interaction* of these bodies.

On the other hand, the peculiarities in the structure of the solar system that were discovered by Copernicus showed that the sun, being at the centre of the system of planets, played a definite role in their motion. In assuming that the sun and planets possess the property of attracting, Copernicus recognized the influence of the sun on the planets. The part played by the sun in planetary motion was still more clear-cut in Kepler's Laws. The sun was in a focus of all the elliptical orbits of the planets (First Law); the planets moved faster when closer to the sun and slowed down when the distance increased (Second Law); the orbital periods of all the planets and their solar distances are related by a single regularity (Third Law).

Kepler himself believed that planetary motion was governed by the sun. He propounded correct views on gravitation, proclaiming that "two separate bodies strive towards each other like two magnets"; gravitation, according to Kepler, has great importance in planetary motion. It holds the planets to the sun. However, Kepler's views on this question were not exactly correct. He believed that the sun attracted the planets like a magnet and in its rotation pulled the planets around.

The next step in the development of conceptions concerning the relation between planetary attraction and motion is closely connected with discoveries in the mechanics of material bodies.

Science is indebted to Galileo for these discoveries. Galileo's work upset the erroneous views on the motion of material bodies that had dominated the world for two millennia, and lay the foundation of a mechanics that is in use to this day.

Before Galileo the dominant conception was that bodies could move only as long as a force acted on them, and in the absence of forces, bodies should remain at rest. This point of view had seemed so in accord with the facts of everyday life that it had remained unquestioned for ages. It was the genius of Galileo that first found it to be fallacious.

Galileo carried out a number of experiments with a smooth and polished metal ball rolled down a smooth inclined board. If the ball is pushed upwards it slows down, stops and then begins to descend at an accelerated rate. Galileo found that if the board is made less sloping the ball will be less retarded in its upward movement and less accelerated in its descent.

And what if the board is exactly horizontal? Obviously, the ball should then experience neither retardation nor acceleration and should roll for an indefinitely long period of time without accelerating or decelerating its motion. In Galileo's own words: "When a body moves in a horizontal plane without encountering any resistance . . . this motion is uniform and would continue without end if the plane extended out into space without end?"

Of course, practically, such an experiment with a horizontal board is impossible, for the ball would stop in any event due to the friction of the board. This is precisely where Galileo's genius comes in, for he was able to abstract himself from the action of friction and to perceive that bodies are accelerated or decelerated due to the action of different forces. And if the body is not acted upon by forces it will move at an unaltered rate (uniformly) and in a straight line. This property of a body to move in the absence of an acting force uniformly and rectilinearly became known as *inertia*.

Galileo's discovery permitted an entirely different view of the causes of planetary motion. To explain the motions of the planets Kepler required the concept of a "pushing" force. This was now unnecessary since the planets could move without it, by inertia. The forces acting on the planets were invoked to explain not the fact that they moved but only the peculiarities of the motion. It was this approach that made possible a correctly established relationship between attraction and planetary motion.

In 1666 the Italian scholar Borelli said the following on the role of gravitation:

"Let us suppose that a planet tends toward the sun and at the same time, in its circular motion, recedes from this central body in the middle of the circle. If these opposing forces are equal they will balance, and the planets will continue to revolve about the sun."

The English scientist Robert Hooke went further. In 1674 he wrote in his paper "An Attempt to Prove the Motion of the Earth from Observations":

"[At a future date] I shall explain a System of the World differing in many particulars from any yet known, [and] answering in all things to the common rules of mechanical motions. This depends upon three suppositions: first, that all celestial bodies whatsoever have an attraction or gravitating power towards their own centres. . . . The second supposition is this: that all bodies whatsoever that are put into a direct and simple motion, will so continue to move forward in a straight line, till they are by some other effectual powers deflected and bent into a motion, describing a circle, ellipse, or some other more compounded curve line. The third supposition is: that these attractive powers are so much the more powerful in operating, by how much the nearer the body wrought upon is to their own centres."

Borelli and Hooke were now not far from the truth. But their ideas were mere conjectures. What was needed was rigorous *proof* that the planetary motions obey the forces of attraction and that the existence of gravitation really explains the observed regularities of these motions.

This was done by the great English scientist Isaac Newton (1642-1727).

Newton became engaged in problems of gravitation and planetary motion as early as 1665-66 (almost at the same time as Borelli and Hooke) and by 1680 he already had a complete theory of his own. Newton published the results in 1687 in his famous book *Philosophiae Naturalis Principia Mathematica* (*The Mathematical Principles of Natural Philosophy*).

In this remarkable work Newton embraced in a single summarization all the previous discoveries in the field of terrestrial and celestial motion and formulated his famous "Laws" that formed the basis of mechanics—one of the theoretical fundamentals of technical science.

Newton's First Law (the law of inertia) states that every body continues in its state of rest, or of uniform motion in a straight line, unless it is compelled to change that state by forces impressed upon it.

Newton's Second Law relates the acceleration of a body to the impressed force. According to this Law the acceleration, w, acquired by a body is proportional to the acting force, F, and inversely proportional to the mass of the body, m:

$$w = \frac{F}{m}$$

In this Law we encounter the concept of *mass*, which was first derived by Newton. Newton calls mass the measure of the quantity of matter contained within a body.

Newton's Third Law states that to every action there is an equal and opposite reaction.

These three laws are fully borne out by practice in terrestrial conditions and constitute the basis for studying the motions of material bodies on earth. Newton applied them to the motions of celestial bodies not doubting for a moment that celestial bodies are subject to the same laws as are terrestrial.

Fig. 18 is, schematically, a small part of the path of a planet, P, round the sun, S. P_0 is the position of the planet at a certain instant of time. If no forces were acting on the planet it would move, by the law of inertia, uniformly and in a straight line with the speed it had at point P_0 (the direction of the planet's motion is indicated by an arrow). This motion would be tangent to the curve at this point. But since the planet moves in a curve it is acted upon by some force that compels it to deviate from a rectilinear path. A planet's orbit is always concave towards the sun, that is to say, the deviation of a planet from a rectilinear path is always directed towards the sun. For this reason, the force acting on the planet should be directed towards the sun.

Newton proved that if the motion of a material body about a certain centre satisfies the Law of Areas, then the force that deflects the material body from a straight

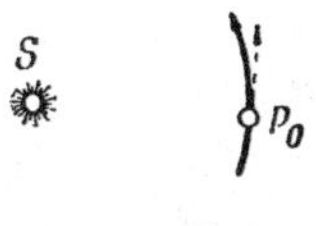

Fig. 18. How a planet deviates from a straight path under the action of solar attraction

line is not merely in the general direction of the centre but always exactly towards the centre.*

Since the motion of planets round the sun satisfies the Law of Areas, these planets should move about the sun due to attraction toward the latter. Thus, what formerly was propounded as a speculation now became a rigorously proven fact.

Newton then proceeded to derive the formulae that permit determining—on the basis of the geometric properties of the curve described by a material body in its motion about a centre of force—the law of variation of the force of attraction with distance from this centre.

He based his conclusions on the fact that the curve in which a body moves will deviate the greater from a straight line (that is, the curvature will be the greater), the stronger the attraction of the centre. Newton's calculations showed that if a curve described by a material body is an ellipse, in one of the foci of which is the attracting centre, the force of attraction of this centre will diminish in proportion to the square of the distance from it.

From Kepler's First Law we know that each planet moves in an ellipse with the sun at one of the foci. Therefore, the force of attraction acting on a planet is inversely proportional to the square of the distance of the planet from the sun. Thus, starting with Kepler's planetary geometry, Newton was able to prove that the planets move due to solar attraction, which varies in inverse proportion to the square of the distance of the planet from the sun.

Planetary motion about the sun was now explained. But Newton reasoned that the force of attraction should also govern the movements of such bodies as the satellites of planets. As Newton saw it, the satellites too should be attracted to their primaries with a force that varies inversely with the square of the distance from the centre of the planet.

When Newton entered the scene four of Jupiter's satellites had been discovered by Galileo and five of Saturn's had been discovered during the period between 1655 and 1684 by Huygens and Cassini. Observations of the Jovian

---

* A rigorous and detailed proof of this and others of Newton's theorems is given in the Appendix.

moons showed that they move uniformly in circles with Jupiter in the centre of their orbits. The movements of these satellites satisfied exactly Kepler's Third Law (the squares of the periods of revolution of the different satellites vary as the cubes of their distances from Jupiter), and this ratio held for the satellites of Saturn too.*

Insofar as uniform circular motion obviously satisfies the Law of Areas, from Newton's theorems it followed that the forces, which compel the satellites of Jupiter and Saturn to deviate continually from a rectilinear path and trace out circles, are directed towards Jupiter and Saturn. Newton proved a simple theorem which stated that material bodies moving around a centre of force in circles and satisfying Kepler's Third Law are attracted to the centre with a force that varies inversely with the square of the distance. From this it follows that the force with which the satellites of Jupiter or Saturn are attracted to their primaries is inversely proportional to the square of the distance from the planet.

Newton thus gave rigorous proof that all the motions of bodies in the solar system result from the action of the force of gravity.

But this was not all. Newton found yet another, indirect, proof of the law of gravitation that he had discovered. Let us reason along with the discoverer. Take a material body moving under the action of the attraction of a centre of force which obeys the inverse-square law. What regularities should we then observe in the motion of this body?

Newton proved the following theorem: if a material body is in motion due to the attraction of a centre of force, this motion should satisfy the Law of Areas. If this force of attraction diminishes with the square of the distance from the centre, the body can move in one of the following curves: ellipse (a special case of which is the circle), parabola or hyperbola. Now if several material bodies are moving in different ellipses about an attractive centre, then, given a force of attraction towards the centre that is in-

---

* Actually, the satellites of Jupiter and Saturn move not in circles but in ellipses of very small eccentricity (observations in those times were not accurate enough to detect ellipticity in their orbits). However, this in no way keeps them from obeying Kepler's Third Law.

versely proportional to the square of the distance to it, the squares of the periods of revolution will vary as the cubes of the semi-major axes of their elliptical orbits.

Summarizing, if the planets move due to gravitational forces then Kepler's three laws must hold. And since these laws were deduced from observations Newton's reasoning served as indirect confirmation of the existence of gravitational forces.

But Newton was not only able to prove that the existence of attracting forces between the planets and the sun follows from Kepler's laws. He succeeded in linking up the gravitational forges, which appeared to act only between celestial bodies, with such a familiar occurrence as the falling of bodies to the earth. Newton studied the lunar motions and proved that the force with which the moon is attracted to the earth and which governs the moon's movements is nothing other than the force of gravity on earth, which force extends to the moon losing strength with the square of the distance from the earth.

Galileo had already conducted experiments which showed that heavy bodies fall to earth with a uniformly accelerated motion equal to roughly 9.8 m/sec^2. And what is more, all heavy bodies moving in any direction (up, down, or at an angle to the horizon) possess the same acceleration directed vertically downwards. Hence, according to Newton's Second Law, all bodies are acted upon by a force directed vertically downwards (toward the centre of the earth), which is the same as to say that all bodies are *attracted* to the centre of the earth. And this is the force that determines the gravity (weight) of the bodies. Bodies possess weight both at the earth's surface and at high-mountain altitude. And so what is there to stop this earth-centred attractive force from operating at a greater distance from the surface? Newton naturally suspected it to extend out at least to the moon. And this force diminished, so Newton reasoned, according to the same inverse-square law that was evident in the case of the sun, Jupiter, and Saturn. Newton was able to verify his conjecture by calculation since at that time both the moon-earth distance and the earth's radius were known.

At the earth's surface, that is, at a distance of about 6,370 km. from its centre, the rate of acceleration is roughly

9.8 m/sec². Acceleration produced by the earth's gravitation at the lunar distance, r, (about 384,000 km.) should diminish by a factor of $\frac{(384,000)^2}{(6,370)^2} \approx 3,640$. Dividing 9.8 m/sec² by 3,640 we obtain 0.270 cm/sec², This should be the acceleration of the moon if it is caused by the earth's attraction. Now let us compute approximately the actual acceleration of the moon, on the assumption that it moves in a circle. In uniform circular motion, the acceleration, w, is equal to

$$w = \frac{v^2}{r}$$

where v is the speed of the moon in its orbit, while r is the distance between the moon and the earth. The speed, v, of the moon in its orbit is equal to $2\pi r/T$, where T is the orbital period of the moon, equal to 27.33 days. Solving the equation we obtain

$$v \approx 1.02 \text{ km/sec.}$$

Squaring v and dividing by r, we find

$$w \approx 0.271 \text{ cm/sec}^2.$$

Thus, the actual acceleration of the moon nearly coincides with that caused by the earth's attraction. The agreement between these two numbers (0.270 and 0.271) is very good if, in addition, we take into account that the moon's orbit is not an exact circle but an ellipse.

Consequently, weight at the surface of the earth and the motion of the moon are due to the same force. This force with which the earth attracts all material bodies to its centre is inversely proportional to the square of the distance from the centre. Thus, weight on earth is the force with which this planet attracts these bodies.

Newton clinches this argument with the following reasoning. Suppose the earth like Saturn or Jupiter were circled by several moons (satellites). The attracting force that holds them in their orbits is inversely proportional to the square of the distance from the earth's centre. Now if the closest satellite were so close as to almost touch the tops of the highest terrestrial mountains, the attracting force that would be maintaining it in orbit would be some 3,640 times that acting on the moon. The acceleration of this satellite would then be roughly

$$0.270 \times 3,640 \ \text{cm/sec}^2 \approx 9.8 \ \text{m/sec}^2.$$

Like all material bodies, this satellite would have weight with the concomitant acceleration of 9.8 m/sec^2 towards the centre of the earth. If the force of gravity differed from the force that held the satellite in its orbit, its acceleration would be the sum of the two accelerations (one due to the force of gravity and the other to the force governing the motion of the satellite), in other words

$$9.8 + 9.8 = 19.6 \ \text{m/sec}^2.$$

But since everywhere on earth the acceleration of falling bodies is equal not to 19.6 m/sec^2 but to 9.8 m/sec^2, the force that holds the moon in its orbit is the same force that we on earth call weight.

The foregoing suggests the conclusion that the revolution of all planets about the sun, the Jovian satellites about Jupiter, Saturn's satellites round Saturn, and the moon orbiting the earth are phenomena of the same nature. All satellites and planets receive their motion from a force directed to the centre of the body around which they move. This force diminishes as one recedes from Jupiter, Saturn, the sun or earth in proportion to the square of the distance. Consequently, these bodies possess the property of attracting other celestial objects. On earth, this attraction embraces all material bodies producing what is known as weight. It seems natural, therefore, that material bodies on the sun, on Jupiter, Saturn, Venus, Mars and Mercury should have the same property—weight.

As follows from Newton's Third Law, attraction is a mutual property. Therefore, if the sun draws to itself all the planets, then each planet should draw to itself the sun. If the earth attracts the moon, the latter should attract the earth. And, finally, if all material bodies are drawn towards the earth, they should also draw the earth to themselves. This suggests that the property of gravitation is innate not only to every celestial body, but in general to all material bodies, all material particles that comprise these bodies.

On what does the magnitude of the attracting force depend? We already know that this force decreases with the distance between the gravitating bodies. But what else is there to alter the magnitude of this force?

Newton proved very simply that the gravitational force is a function of the mass of the body: the greater the mass the stronger the pull it exerts on other bodies.

Experiments on the earth show that the force with which the earth attracts all material bodies imparts to them all the same acceleration (9.8 m/sec^2). Newton's Second Law states that the acceleration $w = F/m$. If the acceleration of all falling bodies is constant, then the force acting on a body should vary in proportion to the mass of the body, increasing and diminishing with the mass.

Or we could argue this way. The attraction of material bodies by the earth determines their weight, which increases in proportion to the quantity of matter contained in them. Hence, the attracting force is proportional to the masses of the bodies.

The force of attraction is thus proportional to the mass of the attracting body. For instance, if there are three bodies: A, B, and C of masses m_A, m_B, m_C, body A will attract bodies B and C with a force proportional to their masses:

$$\frac{F_{AB}}{F_{AC}} = \frac{m_B}{m_C}.$$

But Newton's Third Law states that a body attracted by another body should itself attract the latter with the same force. Consequently, bodies B and C attract body A with forces equal to F_{AB} and F_{AC} and proportional to the masses of these bodies. Hence, the forces of attraction are proportional to the masses of the attracting bodies.

It was from such reasoning that Newton arrived at his famous *law of universal gravitation*:

every particle of matter attracts every other particle with a force directly proportional to the masses of both particles (to the product of their masses) and inversely proportional to the square of their distance apart.

Mathematically, this law may be expressed as follows. Denoting the masses of the material particles by m_1 and m_2, the distance between them by r, we find the force of gravitation F to be:

$$F = f\frac{m_1\,m_2}{r^2}$$

The number f is called the *gravitation constant* and is the same for all material particles. At the present time, the gravitational constants have been determined with a sufficient degree of accuracy. If we take the solar mass as unity, the mean sun-earth distance as unit distance and the mean solar day as unit time, then $f = 0.000295912$.

In c.g.s. units (centimetre, gram, second), f is equal to 6.67×10^{-8} accurate to 0.01×10^{-8}. The possible error here is 0.005×10^{-8}.

4. THE ATTRACTION OF MATERIAL BODIES OF DIFFERENT SHAPES

The inverse-square law was formulated by Newton for material particles. But celestial bodies—the sun, moon, planets—are not material particles. The natural question is: can we study the motions of these bodies by applying Newton's law? What law states how the attraction of material *bodies* varies with distance?

We consider two material bodies A and B whose dimensions are very small in comparison to the distance between them. Mentally, we divide bodies A and B into a large number of very small parts which we shall call "particles." The particles of the bodies attract each other in keeping with Newton's law, and the total attraction of A and B builds up from the mutual attractions of the individual particles. But all the particles of A are, practically, at one and the same distance from the particles of B. Therefore the resultant attractive force of the particles of bodies A and B will be inversely proportional to the square of the distance between these bodies. Thus, Newton's law holds for material bodies whose dimensions are very small as compared to the distance between them. In mechanics these bodies are termed *material particles*.

Material particles are sometimes regarded as small-sized bodies. However, in some problems even the sun and planets may be regarded as particles. Indeed, let us consider the problem of the motion of planets about the sun. Here, the distances of the planets from the sun are great in comparison to their sizes (the earth-sun distance is roughly 100 times the solar diameter and 10,000 times the earth's diameter), which means that, without committing a gross error, the sun and planets may be regarded as attracting each other like particles.

However, by no means in all cases is the distance between the attracting bodies great.

Take, for instance, the earth's pull on a small material body close to the surface. In this case, the very concept of distance between the earth and the body becomes vague, since one may speak of the distance to the earth's surface or to the centre of the earth, etc. If we mentally divide the earth into small particles of equal mass, these particles will attract our body with *different* forces.

Then what law will describe the overall total attraction of the earth if each particle of the earth follows the inverse-square law in pulling the body to it?

When we compared the force of gravity at the earth's surface and the force that holds the moon to the earth, we considered that the earth attracts all material bodies both close to its surface and at a large distance with a force inversely proportional to the square of the distance from the centre of the earth. Isn't there some contradiction here?

This problem was first posed and solved by Newton. He proved a theorem according to which a uniform sphere consisting of particles that attract by the inverse-square law attractsother material bodies and is attracted by them as a single material particle localized in the centre of the sphere and concentrating within itself the entire mass of the sphere (see Appendix for the proof). This remarkable theorem is valid not only for a uniform sphere but also for a sphere whose density varies only with the distance from the centre of the sphere.

This theorem holds for the sun, earth and other planets since they all are nearly spherical in shape. Thus, the sun and planets attract each other like material points for two reasons:

1) the distance between them is very great in comparison with their dimensions;

2) they are nearly spherical in shape.

Of course, not all bodies are spherical and not in all cases are the distances between the attracting bodies great in

comparison to their sizes. For
them the law of force varying with distance becomes
more complicated.

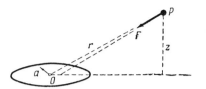

Consider, for example, the
attraction of a particle, P,
(Fig. 19) by a material annulus of mass M and radius a,
the thickness of which is very
small in comparison with the

Fig. 19. The force with which
an annulus attracts point P does
not pass through its centre O

radius. Let particle P be at a distance r from the centre
of the annulus O and at a distance z from the plane of the
annulus. The particles of the annulus closest to P will
attract it with greater force than those farther away, with
the result that the direction of the gravitating force of
the annulus, F, will not pass through the centre of the
annulus but will be displaced towards the half closest to
P. To find the magnitude of the force F, we break the
annulus into tiny ("elementary") particles, each of which
attracts point P according to Newton's law. By summing
all these elementary forces we obtain the force of attraction
of the entire annulus.

Qualitative calculations show that the force of attraction by the annulus of particle P of unit mass at a comparatively great distance may be expressed approximately by
the equation:

$$F = fM \left(\frac{1}{r^2} + \frac{3}{4} \frac{a^2}{r^4} - \frac{9}{4} \frac{a^2 z^2}{r^6} \right)$$

From this equation it may be seen that the attraction of
the annulus differs from that of a sphere of the same mass
with its centre at O by

$$fM \left(\frac{3}{4} \frac{a^2}{r^4} - \frac{9}{4} \frac{a^2 z^2}{r^6} \right)$$

which varies in complicated fashion with the distance r and
the position of the particle P relative to the plane of the
annulus.

Of special interest is the law of attraction of an *ellipsoid
of revolution*, which may be generated by revolving an
ellipse about its minor axis. In this figure, shown schematically in Fig. 20, the distance OL is less than that of OE;

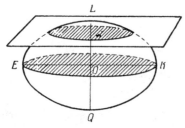

Fig. 20. An ellipsoid of revolution

the plane cutting the axis LQ is an ellipse with semi-axes OL and OE, while any plane cutting perpendicular LQ is a circle. The axis LQ is called the *axis of rotation* of the ellipsoid, the section EK is called the *equatorial section*, its radius OE, the *equatorial radius*, and the distance OL, the *polar radius*. The quantity $\frac{OE-OL}{OE}$, that is, the difference between the equatorial and polar radii, expressed in fractions of the equatorial radius, is known as the *oblateness* of the ellipsoid. An ellipsoid of revolution with small oblateness (differing but slightly from a sphere) is often termed a *spheroid*.

Compared to a uniform sphere, a uniform spheroid of radius OL has excess mass concentrated chiefly along the equator (Fig. 21). The attraction of this surplus equatorial mass should resemble that of an annulus. Therefore, the difference in the attraction of a sphere and a spheroid of equal mass should be roughly the same as in the case of a sphere and an annulus. The force will not vary exactly in proportion to the square of the distance from the centre of the spheroid. It does not pass exactly through the centre but is displaced towards the half of the equatorial section of the spheroid closest to the attracting point. True, since the spheroid is symmetrical with respect to the axis of rotation, the force of attraction will pass through this axis.

It is possible to compute the expression for the force of attraction of a homogeneous spheroid. If a particle of mass $m_1=1$ is relatively distant from the spheroid, that is to say, the distance r from the centre of the spheroid is far greater than the equatorial radius of the spheroid a, we have, approximately,

$$F = fM\left[\frac{1}{r^2} + \left(\frac{3}{5}\frac{a^2}{r^4} - \frac{9}{5}\frac{a^2 z^2}{r^6}\right)\varepsilon\right],$$

where ε is the oblateness of the spheroid, M is its mass, a the equatorial radius, and z the distance of the particle from the plane of the equatorial section of the spheroid.

The force of attraction of the spheroid differs from that of a sphere by the following amount:

$$F' = fM \left(\frac{3}{5} \frac{a^2}{r^4} - \frac{9}{5} \frac{a^2 z^2}{r^6} \right) \varepsilon.$$

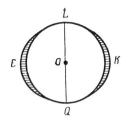

Fig. 21. The excess mass of a spheroid is localized at its equator

This expression is very similar to the difference between the forces of attraction of an annulus and sphere.

The fact that the force with which a spheroid attracts (and is attracted) does not pass through its centre permits of drawing the following important conclusion:

a material particle P, acting on a spheroid with a force F, not only imparts to the spheroid translational motion (determined by the motion of the centre of the spheroid, O), but also tends to turn the axis of rotation of the spheroid LQ.

This is clearly seen in Fig. 22. Particle P attracts both the centre of the spheroid and its equatorial bulge. But the attraction of the closer part of the spheroid is greater than that of the more distant part since $PK < PE$; particle P not only attracts the spheroid imparting translational motion but also tends to turn the equatorial plane EK in the direction of the particle OP.

Considering the expression for the attractive forces of an annulus and spheroid, it may be remarked that as the attracting particle recedes, that is, as the ratio a/r diminishes, the difference between the attraction of these bodies and that of a sphere will decrease and, if r is very great in comparison with a, the annulus or spheroid will exert a force that practically coincides with that of a sphere. As we have seen, this holds for bodies of any shape. At very great distances they all attract (and are attracted) like spheres or, to be more precise, like material particles of the same mass localized in the centres of gravity of these bodies.

The above detailed examination of the attraction of ellipsoids was necessary because the earth and the other planets are slightly oblate spheroids. The earth, for instance, has a polar radius smaller than the equatorial radius by 21 km.,* for Jupiter the difference is 9,500 km. Accordingly, the earth's compression amounts to 1/297, Jupiter's to 1/16, while the greatest of all in the solar system is that of Saturn—nearly 1/11.

Still, the effect of the compression of the planets on solar gravitation is very small. It may be computed that

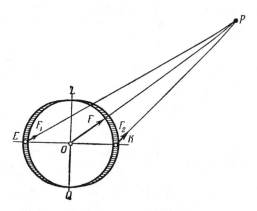

Fig. 22. The attraction of a spheroid

the solar attraction of the earth differs from that of a sphere of terrestrial mass by not more than one part in 150,000,000,000. Thus, in considering planetary motion about the sun one may totally ignore oblateness (that is, one may regard these bodies as material points). However, there are a number of problems in which one has to take into account the supplementary attraction of the planet caused by the departure from spherical shape.

5. EXPERIMENTAL DETECTION OF ATTRACTION BETWEEN MATERIAL BODIES ON EARTH

The mutual attraction of material bodies was first detected in the heavens. But Newton's law applies to all material particles irrespective of where they are located, and therefore attraction should exist between terrestrial bodies as

* The equatorial radius of the earth, accurate to within one kilometre, is 6,378 km., the polar radius—6,357 km.

well. This was actually found to be
the case 50 years after Newton's
discovery, in the eighteenth century. During a scientific expedition to South America in 1735-38
the French scientists Bouguer and

La Condamine noticed that in *Fig. 23.* The deviation of
the vicinity of mountains a plumb plumb bob near a mountain
bob is deflected towards these
mountains. In 1774 the English scientist Maskelyne
made a very careful measurement of the deviations of a
plumb bob. He compared the direction of a plumb bob on
both sides of a narrow mountain ridge Schiehallion in Perth-
shire, Scotland, and found that these directions differ by
roughly 24'', which means that the attraction of the mountain deflected the plumb bob about 12''. One can very
simply determine the force required to deflect 12'' a plumb
bob at the end of a line. Correlating this force with the weight
of the bob, one can find out how much greater is the attraction of the earth than that of the mountain. More, from
the dimensions of the mountain and its surmised density
it is possible to evaluate the mass of the mountain and
then also the mass of the earth as a whole (say, in grams
or tons).

This first determination of the absolute mass of the
earth was still inaccurate since the mass of the mountain
had not been evaluated with any precision. A British scientist, Henry Cavendish, conducted more accurate experiments in 1798 and measured the attraction exerted on a
small ball not by a mountain but by heavy lead balls, whose
masses were accurately known. In these measurements he used
a torsion balance, whose main component is a thin horizontal
rod with small balls on the ends suspended by a slender
elastic fibre (Fig. 24). If massive balls of lead are placed
close to the small balls the latter will be attracted and the
whole rod will turn and thus twist the suspension. The
force acting on the small balls is determined from the
torque. By comparing this force with the weight of the balls,
that is, with the attractive force of the earth, it is possible
to calculate how many times the mass of the earth is greater than the masses of the large balls. Cavendish found the
mass of the earth to be about 6×10^{27} grams $= 6 \times 10^{21}$ tons.

47

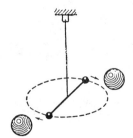

Fig. 24. The Cavendish experiment

Since that time, numerous experiments have been carried out to correlate the force of gravity with the attractive force of terrestrial bodies. These experiments have resulted rather confidently in a more precise value for the mass of the earth— 5.974×10^{27} grams.

If we know the mass of the earth it is possible to calculate the force of attraction between any material bodies. By way of illustration, let us find out the force with which two balls of mass one kilogram each at 10 cm. apart attract each other.

Supposing the radius of the earth to be about 6,000 km., or 6×10^8 cm., the force of attraction between these two balls is less than that of the balls to the earth by a factor of

$$\left(\frac{10}{6 \times 10^8}\right)^2 \times \frac{6 \times 10^{21}}{0.001} \approx 1,600,000,000.$$

The earth attracts these balls with a force of one kilogram, while their mutual attraction is beyond all comparison and amounts to about six ten-millionths of a gram.

Similarly, one can calculate that two ships, each with a displacement of 10,000 tons, passing at a distance of 100 metres will be attracted to each other with a force of 60 grams.

The foregoing attests to the smallness of the forces of attraction between bodies on earth and to the reason why in our daily lives we do not perceive the action of these forces.

But it would be misleading to think that the law of universal gravitation has no practical application in our earthly doings. The point is that the acceleration of gravity at the earth's surface, due to the force of attraction of the earth, is not everywhere constant. It varies with the distance of the attracted particle from the centre of the earth. Since the earth is oblate (a flattening at the poles) and its equatorial radius is greater than the polar radius, acceleration due to gravity on the equator is slightly less than at the poles.* Gravitational acceleration may be measured at different altitudes above sea level. The findings of such measurements made in different latitudes and longitudes enable us to determine variations in the distance to the earth's centre, and, in this way, to find the exact shape of our planet.

However, if we take into account the height above sea level and variations associated with the shape of the earth, acceleration due to gravity is found, in certain localities, to experience additional deviations. These are what are known as *anomalies* of gravity. They result from the upper layers of the earth's crust being inhomogeneous. These layers may contain heavy rocks in some places and light rocks in others. In the first case the gravitational acceleration is slightly above average, while in the second it is somewhat below average for the given locality. The magnitude and character of anomalies due to gravity are a clue to the distribution of rocks of different density in the earth's crust. One is able to "peep" into the earth's interior and get a picture of its geological structure. This helps in geological prospecting for mineral resources.

The science that deals with measurements of variations in acceleration due to gravity is called *gravimetry*. The origin of this science shows how a law discovered in the "sky" finds wide application in the study of purely terrestrial problems. This illustrates vividly that there is no hard and fast dividing line between the "terrestrial" and the "celestial," that nature is integral, with the same laws operative throughout the universe.

6. NEWTON'S LAW—THE THEORETICAL BASIS OF CELESTIAL MOTION

The laws of mechanics and the law of gravitation that Newton discovered not only explained the apparent motions of the planets but opened up to astronomy entirely new vistas. This was the foundation of celestial mechan-

---

* At the equator, acceleration due to gravity is also diminished because of the action of a centrifugal force, which at the poles is zero and on the equator a maximum.

ics, the science that treats of the motions of heavenly bodies.

Formerly, the sole object was to derive the true motions of bodies in space from their apparent paths in the heavens, and then to describe these actual motions by means of some geometric conception. The study of motions was thus of a geometric and descriptive nature.

After the discovery of the law of universal gravitation that governs the motions of celestial bodies, theoretical investigation of motions superseded the descriptive method. It was now possible to solve the theoretical problem of the motion in space of bodies under the action of mutually attractive forces. The past and future motions of celestial bodies were no longer a matter of guesswork and deductions by analogy, but the subject of mathematical calculation.

The first problems of the motion of celestial objects considered as ordinary material bodies subject to mutually gravitating forces were solved by Newton himself. One of the simplest, and yet most basic is the problem of the motions of two bodies attracting each other by the Newton's law (the so-called "two-body problem").

We have already considered the problem of determining the force from a given motion. We said that if a material body is in motion along an ellipse and about a centre of force at a focus of the ellipse, this motion should be due to the attractive force of the centre and should vary in accord with the inverse-square law. The present problem is just the reverse — determine the motion of a material body from a given force. We need to find out how a material body attracted to a centre of force, C, will move.

Fig. 25 shows a centre of force C attracting (according to Newton's law) like a material point of mass M, and a body, B, moving under the action of the attraction of the centre of force, C. The force with which C draws B towards it is:

$$F = f \frac{M m}{r^2}$$

where r is the distance between C and B, m is the mass of B, and f is the constant of gravitation. The quantity

fM defines what Newton calls the "absolute force of the centre."

Let the body B occupy position B_0 at a certain instant of time t_0 which we shall call the *initial instant*. The body's subsequent motion depends upon the velocity which it has at the initial instant. Let this *initial velocity* be v_0 and directed as indicated in Fig.25. If B were not acted upon by any other force it would continue in uniform

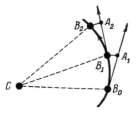

Fig. 25. The path of a body continually curves due to gravitation

and rectilinear motion with the same velocity, v_0, and during a small interval of time Δt it would cover the path B_0A_1. But the force of attraction of C deflects the body from its straight path and during this time it will describe a small arc B_0B_1. B_1A_1 is an indication of deflection from a straight line; its length is connected with the magnitude of the force of attraction. When we determined the force from the given motion of the body, the arc B_0B_1 and the distance B_1A_1 were known, and we had to determine the force from this distance. In this new problem, however, the force is given and we must find B_1A_1 and then the position of the body, B, on the curve. Since the acceleration, that is, the change in velocity of the body B as a result of the gravitation of C, is known, we can also find the velocity the body will have when it arrives at point B_1. Reasoning in the same way we can find the position of B_2 after the next interval of time Δt, etc.

It is thus possible to find points B_1, B_2, etc., that define the path of the body B.

The nature of the motion of the body will change if the initial velocity is different. We consider motion with an initial velocity $v_0{}^*$, which is greater than v_0. During the interval of time Δt, the body will cover (at this speed) a distance $B_0A_1{}^*$ which is greater than B_0A_1 (Fig. 26). But the deviation from rectilinear motion will be the same in both cases inasmuch as the force at C is the same. For this reason, if the small arcs B_0P_1 and $B_0B_1{}^*$ denote the actual distances covered by the bodies in both cases, then B_1A_1 and $B_1{}^*A_1{}^*$ should be equal. Thus, as an inspection of Fig. 26 shows, increased initial velocity diminishes the curvature of the trajectory.

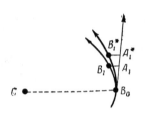

Newton showed that if the initial velocity (for simplicity, we consider $v_0 \perp B_0C$) at a given position of the body (that is, the distance of B from C) does not exceed the value $\sqrt{\dfrac{2fM}{r_0}}$ (where fM is the "absolute force of the centre" and r_0 is the initial distance),

Fig. 26. The greater the initial velocity the less the curvature of the path

the body, B, will describe an ellipse with a focus at C. In the case of small initial velocities, the ellipse will be greatly elongated along the straight line B_0C, with the initial point B_0 the aphelion (Fig. 27). As the initial velocity is increased, the dimensions of the ellipse will increase, as will also the orbital period in this ellipse.

At first, increasing the velocity will make the ellipse rather more expanded than elongated and it will take on a rounded form. At an initial velocity of $v_0 = \sqrt{\dfrac{fM}{r_0}}$ the curve will be a circle. This velocity is known as the circular velocity. Further increases in the initial velocity will produce more and more elongated ellipses with less expansion. Its semi-major axis and the maximum distance at which B recedes from C will increase more and more rapidly. At $v_0 = \sqrt{\dfrac{2fM}{r_0}}$ the curve will no longer be closed. The semi-major axis of the ellipse will have reached infin-

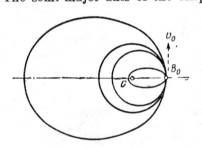

Fig. 27. The orbits of a body moving in a gravitational field with velocities less than critical

ity and the initial velocity will now be so great that the attraction of the centre, C, will no longer be capable of returning B, and the body will recede from C for good. Newton shows that the resulting path will be a *parabola*. Fig. 28 shows the limb of the parabola (middle curve) along which the body B will recede to infinity without returning. (The other limb of the parabola is shown as a dashed line.) The velocity $v_0 = \sqrt{\dfrac{2fM}{r_0}}$ is called the *critical* or *parabolic* velocity.

At $v_0 > \sqrt{\dfrac{2fM}{r_0}}$ the body, P, will even less be able to return to C. As Newton demonstrated, it will move in an open curve called a *hyperbola* (the outermost curve in Fig. 28). The greater the initial velocity v_0, the less curved will this hyperbola be and the faster B will recede from C.

Consequently, the only path a body, B, can have around a gravitating centre is along a conic section: an ellipse (a closed curve), if its initial velocity does not exceed the critical velocity, and a parabola or hyperbola (open curves) if the initial velocity is equal to or greater than critical. In the special case, when the initial velocity is directed normally to $B_0 C$ and its magnitude is exactly equal to $\sqrt{\dfrac{fM}{r_0}}$, B will move about C in a circle. Now if the initial velocity of P is zero, it will simply fall towards C (moving under gravity) along the straight line BC. As B approaches C the force of gravitation, and with it the acceleration of B, will increase.

There is an important relationship between the orbital period of the celestial body, the semimajor axis of the orbit and the mass of the centre of force M:

$$T^2 = \frac{4\pi^2 a^3}{fM}\,.$$

This relationship brings us to Kepler's Third Law (it was precisely in this manner that Newton derived this law from his theorems). Also, we are now able to compare

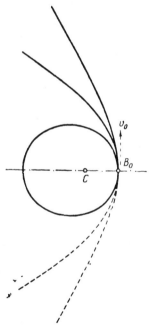

Fig. 28. Ellipse, parabola and hyperbola

the masses of the centres of force about which celestial bodies revolve.

Let us consider the motion of the earth about the sun and the moon round the earth. If we denote the orbital periods of the earth and moon by $T_\oplus$ and $T_\mathbb{C}$, the mean earth-sun and moon-earth distances by $a_\oplus$ and $a_\mathbb{C}$, and the masses of the sun and earth by M and $m_\oplus$, we find that

$$T_\oplus^2 = \frac{4\pi^2\,a^3_\oplus}{fM} \text{ and } T^2_\mathbb{C} = \frac{4\pi^2 a^3_\mathbb{C}}{fm_\oplus}$$

and

$$\frac{T^2_\oplus}{T^2_\mathbb{C}} = \frac{a^3_\oplus}{a^3_\mathbb{C}} \times \frac{m_\oplus}{M}$$

whence

$$\frac{m_\oplus}{M} = \frac{T^2_\oplus}{T^2_\mathbb{C}} \times \frac{a^3_\mathbb{C}}{a^3_\oplus}.$$

Substituting the following approximate values: $T_\oplus = 365$ days, $T_\mathbb{C} = 27$ days, $a_\mathbb{C} = 384,000$ km., $a_\oplus = 150,000,000$ km., we obtain

$$\frac{m_\oplus}{M} = \frac{365^2}{27^2} \times \frac{(384,000)^3}{(150,000,000)^3} \approx \frac{1}{330,000} . *$$

In the same way, one may compare the masses of the sun and any planet that has satellites, or the masses of two planets with satellites. Newton accordingly determined the masses of Jupiter and Saturn whose satellites were known at that time. It turned out that the Jovian mass is less than that of the sun by a factor of 1,000, while for Saturn this figure is 3,000.

These first determinations of the masses of celestial bodies showed that the planetary masses are very small in comparison to that of the sun. When studying the motions of planets around the sun or of satellites around their primaries, use may be made of the results of the solution of the foregoing two-body problem, assuming the sun as the centre of force in one case and the planet as that force in the other. However, the problem of the motion of a planet about the sun differs from the above problem of motion about a centre of force. Indeed, we considered the centre of force stationary, and, hence, did not take into ac-

---

* Newton's own, erroneous determination of the earth's mass at 1/170,000 that of the sun was due to the fact that the earth-sun distance was very imperfectly known at that time.

count the attraction of this same centre of force by the given body. Yet, according to Newton's Third Law each planet should attract the sun, and the sun should move due to the attraction of the planets.

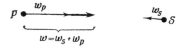

Fig. 29. Calculating the "absolute force of the centre" in the two-body problem

We consider planet P and the sun S attracting each other according to Newton's law. If their masses are m and M, the force of mutual attraction between them will be

$$F = f\frac{Mm}{r^2}$$

where r is the distance PS. The sun, S, imparts to the planet an acceleration $w_p = f\frac{M}{r^2}$, while the planet, drawing S with the same force, imparts to the latter an acceleration $w_s = f\frac{m}{r^2}$. The acceleration of the sun, S, is less than that of the planet, P, the same number of times the mass S is greater than the mass P. The sun's acceleration is obviously very small, but the main thing is that it exists.

Let us consider the movement of a planet P relative to the sun, S, as seen by an observer on the sun. The relative acceleration of the planet P will be equal to the sum of the accelerations w_p and w_s:

$$w = w_s + w_p = f\frac{M + m}{r^2}.$$

Thus, if we regard the motion of P around S, the latter a stationary centre of force, the acceleration from P to S will be such as is created by an attracting body of mass $M + m$. The coefficient that defines the "absolute force of the centre" will be $f(M + m)$. All the equations that define the critical velocity, the orbital period, etc., which are deducible for motion about a centre of force, hold for the new meaning of the "absolute force of a centre." If P revolves about S in an ellipse, then

$$T^2 = \frac{4\pi^2 a^3}{f(M + m)}.$$

If several bodies of mass m_1, m_2,... revolved about a single body S (for instance, the motions of several planets around the sun) with periods T_1, T_2, ... and semi-major axes of their orbits a_1, a_2, ... , then

$$T_1^2 = \frac{4\pi^2 a_1^3}{f(M+m_1)}; \quad T_2^2 = \frac{4\pi^2 a_2^3}{f(M+m_2)}; \quad ...$$

$$T_1^2 : T_2^2 ... = \frac{a_1^3}{M+m_1} : \frac{a_2^3}{M+m_2} ... ,$$

This relation shows that the squares of the orbital periods of the planets about the sun are not exactly proportional to the cubes of the semi-major axes of their elliptical orbits. Strictly speaking, then, Kepler's Third Law is not satisfied. However, since the masses of the planets are very small in comparison with the solar mass, that is, $(M+m_1)$: $(M+m_2)$:.... ≈ 1, then also $\frac{a_1^3}{T_1^2} = \frac{a_2^3}{T_2^2} = ...$ Kepler's Third Law is thus almost exactly satisfied (recall the Table on p. 29).

The motion which we have just examined is *relative motion* of one celestial body about another. But when viewed from outside, from a certain stationary point, both P and C would be found to be in motion due to mutual accelerations. Newton showed that the centre of gravity of C and P would be at rest (or in uniform and rectilinear motion), while the bodies themselves would be moving about this centre of gravity. Their orbits would be similar to the orbit of P in its motion round C.

If the mass M of the centre C is very great as compared to the mass m of body P, the acceleration imparted to C by P is negligible, and the centre of gravity of C and P practically coincides with C. We may then say that C is stationary and that P is in motion about a stationary centre.

Take, for example, the earth's motion about the sun. Since the terrestrial mass is only $1/330{,}000$ that of the solar mass, the centre of gravity of these two bodies is situated at a distance of

$$\frac{150{.}000{.}000}{330{,}000} \approx 500 \text{ km}.$$

from the centre of the sun, which means that the centre of the sun describes around the centre of gravity of the sun and earth a circle of radius 500 km. This is only about 1/2,800 the solar diameter, which is 1,400,000 km. When correlated with the size of the sun and the solar distance, such oscillations in the sun's position are negligible, so that the sun is practically stationary in this sense.

But when computing the earth's orbital period about the sun a knowledge of the terrestrial mass is necessary. If the earth's orbital period is computed from equations $T^2 \sim \dfrac{a^3}{fM}$ and $T^2 \sim \dfrac{a^3}{f(M+m)}$ we get a discrepancy of roughly 100 seconds. This is a rather perceptible quantity.

7. CELESTIAL MOTION AND THE TWO-BODY PROBLEM

The problem of the motion of two bodies attracting each other according to Newton's law is widely used in astronomy. Planetary motion round the sun is first of all regarded as motion due to solar gravitation in accordance with Newton's law. The same procedure is used in elementary studies of the motions of asteroids and comets about the sun and of satellites round their primaries. For planets, asteroids and satellites, elliptical orbits are found that accord with the two-body problem.

The motion of a body in space in an elliptical orbit is fully defined by six quantities which are known as orbital *elements*.

Two elements —the *inclination* and the *longitude of the ascending node*—define the position in space of the plane in which the planet moves.

The inclination is the angle i between the plane of planetary motion and the plane of the ecliptic (the plane of the earth's orbit).

The two points at which the planet intersects the plane of the ecliptic are called *nodes*. Near the ascending node ☊ the planet passes from the southern hemisphere of the sky into the northern hemisphere, near the descending node ☋ —from the northern to the southern hemisphere. The angle ☊ between the direction towards the point in the sky ♈, where the sun is located on the day of the autumnal equinox, and the ascending node of the planetary orbit is called the longitude of the ascending node.

The third element—the distance of perihelion from the node—is the angle ω between the directions towards the ascending node and the perihelion of the orbit. This element is sometimes defined otherwise. We call the *line of nodes* the line along which the planetary plane intersects

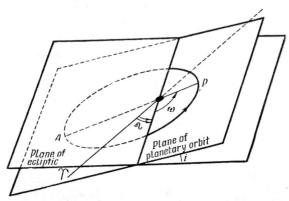

Fig. 30. The orbital elements of a planet

the plane of the ecliptic, and the *line of apsides* that which connects perihelion *P* and aphelion *A*. Then the distance of perihelion from the node will be the angle between the line of nodes and the line of apsides.

The fourth and fifth elements—the *semi-major axis* and the *eccentricity*—define the size and shape of the elliptical orbit of the heavenly body.

And, finally, the sixth element defines the position of the body in the orbit at a certain instant of time, which is commonly the time of passage of the planet through perihelion. If the initial position of a body in orbit is known, then with a knowledge of the mean velocity of motion of this body it is possible, by invoking the Law of Areas, to compute the orbital velocity and, hence, the position of the body in the orbit at any instant of time. If one knows the shape, size and position of the orbit in space, he can determine the position of the given celestial body in space and then compute its apparent position in the sky.

Determinations of the elliptical orbits of Mercury, Venus, Earth, Mars, Jupiter, and Saturn caused no difficulties since these planets had been under observation for centuries and data on their positions in the sky were numerous. These orbits were first derived by Kepler. As time passed, the orbital elements were refined by new and more accurate observations.*

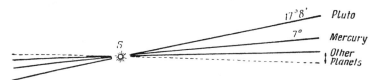

Fig. 31. Orbital inclinations of the planets of the solar system

Table 2, lists several orbital elements (based on modern data) of all the major planets.

We see that the orbits of all the planets, with the exception of Mercury and Pluto, have slight inclinations and eccentricities, that is, they move in practically the same plane (Fig. 31) and are nearly circles.

Table 2

	Semi-major axis		Orbital period	Eccentricity	Inclination
	A. U.**	10^6 km.			
Mercury . . .	0.387	57.9	87.97 days	0.206	7°0'
Venus	0.723	108.1	224.70 days	0.007	3°24'
Earth	1.000	149.5	365.26 days	0.017	—
Mars	1.524	227.8	1 year 322 days	0.093	1°51'
Jupiter	5.203	777.8	11 years 315 days	0.048	1°18'
Saturn	9.539	1426.0	29 years 167 days	0.056	2°29'
Uranus	19.191	2869.1	84 years 7 days	0.047	0°46'
Neptune . . .	30.071	4495.6	164 years 280 days	0.009	1°46'
Pluto	39.457	5898.9	248 years	0.249	17°8'

The two-body problem has also been successful in studying the motions of the minor planets (asteroids), comets,

---

* At the end of the seventeenth century, telescopes began to be used to determine the positions of heavenly bodies, and observational accuracy increased greatly.

** A.U. stands for *astronomical unit*—the mean distance between the earth and sun. It is convenient for measuring distances in the solar system.

and satellites of the major planets. The first application of this problem to asteroids and comets was particularly important as confirmation of Newton's theory of gravitation.

Comets—these "long-haired" luminaries—have been observed since the dawn of man. Prior to Newton, most astronomers believed them to be formations in the terrestrial atmosphere. Galileo and Kepler considered them celestial bodies but could not find an explanation for their strange movements, they could not understand why they appeared suddenly and disappeared without a trace.

Newton argued that since comets were some sort of celestial body they should obey the law of gravitation too. Such being the case, their motions are due to the sun's gravitational pull. But what are their paths? The two-body problem permits of only three types of motion: elliptic, parabolic and hyperbolic. Newton had to exclude elliptical motion more or less close to a circle since comets, obviously, did not circulate regularly about the sun as did the planets. That left the parabola and the hyperbola. Also possible were elongated ellipses, for in such orbits the comets would be visible for only a short time during closest approach to the sun and would then be lost to view for long periods.

The problem arose of determining the path of a comet in space. Observations of all comets were comparatively brief and there were few facts about their position in the sky; all of which made the problem immensely difficult. Newton solved the problem of determining a parabolic orbit from several observations. He showed how it was possible to find the parabola (the elements of a parabolic orbit) of a comet moving in space if three of its positions in the sky are known for different times. Newton applied his method to comets that were observed in 1680 and 1682. He computed the elements of their parabolic orbits, and right up to the time the comets disappeared from sight their computed and observed positions in the heavens were very close. This was proof that comets are celestial bodies moving under solar gravitation. This discovery was one of the first remarkable attainments of the new gravitation theory.

One of Newton's pupils, Halley, used this method to compute the orbits of 24 comets that had appeared chiefly in the sixteenth and seventeenth centuries. Extant records of the positions of these comets served as the starting point. He noticed that the parabolic orbits of comets that had made their appearances in 1682,1607 and 1531 were very much alike. And what is more, the time intervals between apparitions were roughly the same: 1607-1531=76 years, 1682-1607 =75 years. Halley figured that what had actually been observed on these occasions was not three different comets but one and the same comet moving not in a parabola but along a very elongated ellipse. An extremely elongated ellipse nearly coincides with a parabola near the focus, and a comet is, as a rule, observed near the focus at which the sun is located. And it is very difficult to distinguish between motion along a parabola and along an elongated ellipse.

Halley predicted the next appearance of this comet for 1758-59. And true enough, the comet appeared in 1759. This was vivid evidence for the correctness of the law of gravitation.

Gravitational theory was equally successful in connection with the discovery of the minor planets or asteroids.

On January 1, 1801, the Italian astronomer Piazzi detected in his telescope an object that looked very much like a star, but was in rapid motion among the stars. This was similar to the apparent motions of the planets and suggested that the body was relatively close to the sun. Piazzi soon lost sight of this star-like body as it approached the sun and disappeared in the latter's rays. The problem now was to determine the motion of this heavenly body in space on the basis of Piazzi's observational data. It was solved by the German mathematician Gauss who found a way of determining an elliptical orbit from several observations. Gauss demonstrated that if the positions of a body in space are known for three different instants of time, it is possible to define the ellipse along which this body moves. The problem of determining the elliptical orbit from observations proved far more difficult than the problem of determining the elements of a parabolic orbit. But Gauss solved it, and in such rigorous mathematical form that his method of determining the elements of an elliptical orbit is the best to this day.

Gauss computed the orbit on the assumption that the object Piazzi had discovered was in motion about the sun due to gravitational force. It turned out that this heavenly body moved in an ellipse, the semi-major axis of which was roughly 2.8 astronomical units and the eccentricity 0.08, which meant that it had an orbit between those of Mars and Jupiter and that it belonged to the solar system. Gauss

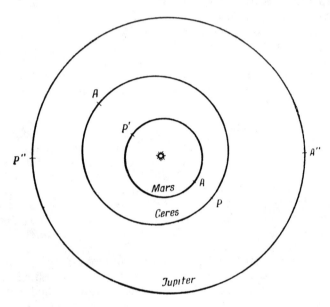

Fig. 32. Ceres' orbit (*A* and *P* denote aphelia and perihelia of the orbits)

computed its positions in the sky for different times thus indicating where it could be observed in the future. And when observations were resumed towards the end of the year (1801) it was recovered in the very spot that Gauss had predicted. In this way the solar system acquired a new planet that was given the name of Ceres.

Three more such objects were soon discovered. In the telescope they appeared like stars, but moved like planets, and so were called *asteroids*—the Greek for "star-like." They are also often called *minor planets* due to their small size in comparison to the *major planets*—Mercury, Venus, Earth, Mars, Jupiter, Saturn, Uranus, Neptune and Pluto.

At present, the elliptical orbits of some 1,600 asteroids have been computed. Each year more and more asteroids are discovered and their orbits derived.

In mass and dimensions the asteroids are far smaller than the major planets. In diameter the largest asteroids reach from 200 to 800 km., while the smallest ones, just detectable in the biggest telescopes, do not exceed a kilometre or two across. Many asteroids have irregular shapes resembling chunks of rock.

8. THE CONCEPT OF PERTURBED MOTION CELESTIAL MECHANICS AND PRACTICAL ASTRONOMY

Descriptions and studies of planetary and cometary orbits based on the two-body problem represent only a first step towards investigating the complex motions of heavenly bodies. Indeed, if a planet, say the earth, were attracted by the sun alone it would pursue a precisely elliptical orbit. Such motion that fits the solution of the two-body problem is known as *unperturbed* motion. But all bodies of the solar system attract each other. So the earth is acted upon not only by the sun but also by the other planets, which "displace" our globe from its elliptical orbit. In turn, the earth attracts the other planets and forces them to deviate from their elliptical paths. A consequence of Newton's law of gravitation is that all the planets, asteroids, and comets of our solar system attract each other and deflect each other from the path that each would pursue under the gravitation of the sun alone. Satellites are constrained to alter their elliptical motion due to the sun's attraction and that of "alien" planets.

Add to this the fact that the planets do not possess true sphericity and do not attract in absolutely exact accordance with the inverse-square law. Though in practice this may be ignored when considering sun-planet interactions or the mutual attractions of the planets (see above, Section 4), satellite theory demands a consideration of these facts. For example, the moon, even in the absence of solar or planetary attraction, would not pursue a precisely elliptical orbit about the earth.

To summarize, then, not a single body in the solar system can describe precisely an ellipse, parabola or hyperbola.

The generic term *perturbations* is given to all these deviations from elliptic, parabolic or hyperbolic motion. Planets, asteroids, satellites, and comets are said to be perturbed. Perturbations are exceedingly involved and to account for them is a task of immense complexity. However, if perturbations are ignored we obtain only a rather crude picture of the motions of many heavenly bodies.

In Kepler's day, when observations to determine the positions of heavenly bodies were made with the naked eye, planetary deviations from an elliptical unperturbed path were not striking. But in the mid-seventeenth century telescopes and various measuring instruments began to be used in this work, and as the astronomical tools became more refined observations became more accurate. Thus it was that the deviations of the planets and of other bodies of our solar system from the Keplerian laws could no longer be ignored.

From the eighteenth century on, one of the principal problems of celestial mechanics was the determination of the perturbations of planets, asteroids, satellites and comets.

Methods for determining perturbations and also techniques for solving other important problems of celestial mechanics grew up together with the methods of higher mathematics. Present-day celestial mechanics was created by the great mathematicians of the eighteenth and nineteenth centuries: Clairaut (1713-1765), d'Alembert (1717-1783), Euler (1707-1783), Lagrange (1736-1813), Laplace (1749-1827), and others. And to this armamentarium of celestial mechanics, the twentieth century added new and powerful mathematical techniques and remarkable calculating machines.

There were two principal reasons for this development of celestial mechanics.

The first was the immediate demands of practical astronomy. Celestial mechanics was being called upon to give solutions to a number of problems of great practical significance. In the eighteenth century (before the invention of chronometers), one such problem was the elaboration of an accurate theory of the moon's motion. At that time, lunar observations were used to determine geographical longitude.

To determine the longitude of a given point on the earth's surface, it is necessary to compare, at a given time, the local time of this place with the local time of some other place whose longitude is known.* The local time at the given point can easily be determined from direct observations of the sun or stars. But how is one, far out at sea, to know what the local time of, say, Greenwich is. True, at the present time special radio time signals are broadcast for use in figuring Greenwich time. In the nineteenth century, before radio had been invented, very precise clocks called chronometers were in use. One could take such a chronometer with him out to sea or on an expedition to a locality with an unknown longitude and thus "take along" Greenwich time. But the eighteenth century had neither radio nor chronometer, and the most accurate clock to find out Greenwich time was the moon.

The moon moves through the stars quite rapidly—roughly 30' per hour. So one can regard the stellar heavens as the clock-face and the moon as the hour hand. The only thing is to graduate this peculiar dial. In this we are aided by specially compiled tables, based on the theory of lunar motion, which indicate the moon's position among the stars at definite moments off Greenwich time. This "celestial time-table" of the moon should, of course, be sufficiently accurate. In order to determine Greenwich time, and thus the longitude, with even such a slight accuracy as one minute (this corresponds to an error of 30 km. in determining a point on earth), it was necessary to know the position of the moon in the sky to within 15". It was exceedingly difficult to create even such a theory of lunar motion because the moon is subject to exceptionally strong perturbations.

At the present time we have more accurate methods of determining longitude. But a precise theory of lunar motion is necessary for other purposes. Without this theory it would be impossible to calculate accurately the beginning of solar eclipses, their duration, to define precisely the locality where the eclipse can be observed, and so on. Such

---

* On the globe, longitude is measured from a meridian that passes through Greenwich, a town near London and the site of the Greenwich Observatory. For this reason, when determining the longitude it is most convenient to know the local time at Greenwich.

predictions are necessary so as to organize properly observations of eclipses. Information about eclipses that were observed in the distant past enable us now to study irregularities in the earth's rotation (see Section 17). Eclipse theory is helping even historians. By correlating the theoretically calculated times and places of eclipses with the written records of these eclipses found in ancient manuscripts it is possible to find out the exact dates of historical events where other dating procedures fail.

The second reason for the development of celestial mechanics was the necessity to verify the law of universal gravitation.

It was the task of celestial mechanics to find out whether Newton's law of attraction was correct, whether the attractive forces between material particles do indeed vary inversely with the square of the distance between them. Is it possible by this law alone to compute all the observed perturbations of the bodies of the solar system? Naturally, there was hardly any doubt that this law gave a good description of the movements of planets, satellites, asteroids and comets. But is it capable of describing accurately *all* the motions of these bodies? Verification of this fact was, of course, of extreme importance both scientifically and practically. To do this, it was necessary to build a theory of the motions of planets and the other bodies of the solar system.

The question of refining Newton's law was frequently raised in connection with calculations of the disturbed motions of celestial bodies. For instance, in the middle of the eighteenth century the French astronomer Clairaut was absolutely unable at first (in building a theory of lunar motion) to correlate all the observed peculiarities of lunar motion with Newton's law of gravitation. He therefore suggested that the force of attraction does not vary in inverse proportion to the square of the distance but is expressed by the following equation:

$$F = \frac{A}{r^2} + \frac{\varepsilon}{r^3}$$

where ε is a very small number.

However, Clairaut later detected inaccuracies in his mathematical procedure that led to an incomplete account of all perturbations in the moon's motion. After the errors were eliminated the disparities between theory and observation disappeared, and there was no longer any necessity for the additional term ε/r^3 in the gravitational equation.

In a number of other cases, for example, in studies of the motion of Mercury, opinions were advanced that the force of attraction varies in inverse proportion to the $(2+\delta)^{\text{th}}$ power of the distance, where δ is a certain very small number. But these were erroneous too. Newton's law described the motion of the planets (with one exception, see Sec. 21) with the same accuracy that present-day observations yield.

Thus, the relationship that Newton established between the forces of gravitation of material particles and their distance apart is now beyond any doubt. Newton's law is borne out by the entire essemblage of observational and theoretical findings.

9. WAYS OF DESCRIBING PERTURBED MOTION. THE VARIATIONAL ORBIT

In the preceding section we pointed out the possibility of regarding perturbations simply as the difference between positions of a body in unperturbed and perturbed motion. But perturbations may be characterized otherwise by applying a method first developed by Euler in 1756.* Underlying this method is the concept of a so-called *variational orbit*.

If the velocity of motion of a body and its position in space is known at a given instant of time, it is then possible to determine the path (ellipse, parabola or hyperbola) which this body will pursue around the sun in the absence of perturbations. Suppose that the elliptical orbit of a planet has been found for a certain instant of time, but under the influence of perturbations the planet deviates from this orbit. If after an interval of time its elliptical orbit is computed from new observations, the result will of course be a different ellipse. True, if the perturbations are small the new ellipse will resemble the preceding one, that is, the new orbital elements will differ but slightly from those derived earlier. If this procedure is repeated some time later, the orbital elements will again differ. For this reason we may

---

* It was published in 1771.

imagine that at each instant the planet is moving along a certain elliptical orbit, but that the orbital elements, that is, the dimensions, shape and position of this ellipse in space, are constantly changing. In this case the planet is said to be moving in a *variational* elliptical orbit.

If a planet or other body does not experience considerable perturbations during a single circuit, its motion during this time will be almost exactly in the same ellipse. But perturbations are cumulative in time. In each subsequent circuit the planet pursues a slightly different ellipse. After a large interval of time, the shape, size and position of the ellipse may be perceptibly altered. For example, the ellipse described by the earth is almost of the same size and shape as 1700-1800 years ago and lies in nearly the same plane. Yet it has turned approximately 5° in this plane. In 1850 Jupiter's semi-major axis was 5.20265 astronomical units, while in 1950 it was 5.20290 astronomical units— an increase of roughly 40,000 kilometres.

Now if the perturbations are great the path of the body will deviate from the elliptical during even a single orbital period. In other words, the orbital elements computed at the beginning and the end of a period of revolution are noticeably different. To illustrate again, take the orbital elements of the moon at the beginning and end of a month. It will be found that during this time the ellipse that the moon should pursue has turned more than 3° in its plane. The lunar elements vary with comparatively great rapidity. But even so the elements of this elliptical variational orbit at a given instant give a pretty fair picture of the motion during a single or several circuits.

Alterations of the elements of a variational orbit are a certain clue to what type of changes are to be expected in the motion of the given body. For instance, an increasing semi-major axis will mean that this body is pursuing an ellipse that is continually increasing in size, that is to say, the body is gradually receding from the central body. If the eccentricity is increasing, the elliptical orbit is elongating—the closest approach of the bodies becoming less and their greatest distance apart ever greater.

Thus, the perturbed motion of a heavenly body can be described by means of the varying orbital elements, that is, those that change with time. The motion of all planets, asteroids, and satellites is perturbed motion, and for each instant of time their orbits are described by specific values of the elements. The table of orbital elements of the major planets on p. 59 is not a list of constant, definitively established magnitudes. These elements refer to our epoch, but since the planetary perturbations are slight the elements change very slowly with time and will describe planetary motion for a long time to come.

10. THE PROBLEM OF MOTION IN THE SOLAR SYSTEM

A precise statement of the problem of motion of bodies of the solar system consists in the following.

The solar system comprises the sun, planets, their satellites, asteroids and comets. All these bodies attract each other in accordance with Newton's law. The problem is to study their motions mathematically when their positions and velocities at a certain (initial) instant of time are known.

There is, of course, no necessity to state the problem in such a general form, with account to be taken of the influence of *all* bodies on each other. The masses of the comets and asteroids are very small in comparison to those of the planets, and so can impart to the sun and planets only absolutely negligible accelerations. The very masses of comets were evaluated on the basis of the fact that they had never been observed to disturb the motion of any planet or of the satellites of any planet. In 1886 a comet passed very close to Jupiter and right in between its many moons. But the latter didn't exhibit the slightest deviations in their ordinary motions. The inference was that the mass of this comet was less than that of the earth by a factor of at least one million.

The masses of the largest asteroids are of course much greater than those of comets. Let us evaluate the mass of the largest asteroid, Ceres, whose diameter is 800 kilometres. If this asteroid has a density that of the earth* its mass should be approximately $(12,800/800)^3 = 4096$ times less than that of the earth. Several asteroids have diameters around 100-300 kilometres, while the majority are much

---

* The earth has a mean density of 5.5 g/cm³ ; true, all the other major planets have a lower mean density.

smaller, so that the mass of all the asteroids together does not exceed 1/700 of the terrestrial mass. Naturally, bodies of so small mass cannot influence perceptibly the motion of planets and satellites.

Consequently, we can consider the motions of planets and their satellites as independent of the motion of asteroids and comets.

To continue, the masses of satellites are but a fraction of the planetary masses. For instance, among the Jovian moons, Ganymede—the third satellite outwards from Jupiter—has the largest mass, roughly 1/12,200 that of its primary, and is 1 million kilometres distant. Under the influence of mutual attraction, Jupiter and Ganymede describe ellipses about their common centre of gravity, which is about 1,000,000/12,200=80 km. from the centre of Jupiter. These relatively very small deviations will have practically no effect on the attraction between the sun and Jupiter. Besides, it is actually impossible to notice these deviations in observations from the earth. Even at the earth's closest approach to Jupiter they cause a displacement in Jupiter's apparent position that does not exceed 1/30″. Still less is the influence of the other Jovian satellites on the motion of their primary. The influence of the satellites of Saturn, Mars, Uranus and Neptune on the motion of their planets is also negligible. An exception is our moon with its mass less than that of the earth by a factor of 81.5 times. The moon and earth move around a common centre of gravity that is 4,700 km. from the centre of the earth. The deviations of the earth from its path by this amount are rather noticeable, particularly in precise studies of the earth's motion round the sun. For instance, the sun-earth direction with account taken of movement around the centre of gravity can differ by 6″ from the direction when no such allowance is made. For this reason, in precise investigations we consider jointly both terrestrial and lunar motion, that is, the entire earth-moon system and the motion of the centre of gravity of this system about the sun.

All this means that instead of considering the motions of the planets, asteroids, satellites and comets jointly, one can pose separate problems of the motion of a) the planets, b) their satellites, c) asteroids, d) comets.

In studies of planetary motion, account is taken of solar

gravitation and the mutual attraction of the planets. Otherwise stated, the problem is that of the motion of 10 bodies (nine planets and the sun) attracting each other according to Newton's law.

When investigating the motions of the satellites of a planet, the latter are considered to occur under the action of the attraction of the planet (main force), the disturbing mutual attraction of the satellites, and also the disturbing attraction of the sun and other planets.

Since the asteroidal and cometary masses are exceedingly small they do not produce any perceptible effect on the motion of the other bodies of the solar system. For this reason, the motion of each comet or asteroid is regarded separately, as if they were attracted by the sun and planets but did not themselves attract.

However, despite such simplifications we always have to deal not with the motions of two bodies but of several gravitating bodies. Mathematically, this problem is so involved that to this day it remains unsolved in its general form. In other words, we are not able to obtain equations which represent the motions of these bodies, that is, which would permit computing the positions of these bodies in space or to get an idea about their properties for *any* masses or initial positions and velocities. And, of course, we are unable to describe these motions in words in a general case, as, for instance, was possible in the two-body problem.

It is, of course, a matter not only of the mathematical complexity of the problem, but also of the extraordinary complexity of the motions themselves even in the case of only *three* bodies. Some idea of these difficulties may be had from a consideration of three bodies: the sun, S, Jupiter, J. and a meteoroid, M, with a mass of several grams. Due to the smallness of its mass, such a meteoroid imparts to Jupiter only the very slightest additional (perturbing) acceleration, and we may regard Jupiter as moving along an elliptical orbit under the influence of the sun's gravitation alone, while the meteoroid's motion is affected by the attraction of the sun and Jupiter. In celestial mechanics, this is known as a *restricted three-body problem*. The restriction here is that the influence of the small body on the other two is ignored, while the motions of these two latter bodies are known.

When the meteoroid is relatively far away from Jupiter the gravitational pull of the planet is but a fraction of that of the sun, and it pursues an elliptical orbit about the sun experiencing only small perturbations. In Fig. 33 M_0 is the initial position of the meteoroid, $M_0 M_1$ is the first portion of the ellipse that it covers. If it reaches M_1 when Jupiter

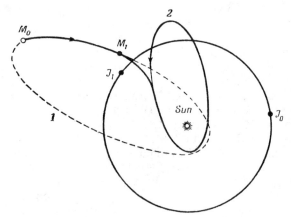

Fig. 33. An example of motion in the restricted problem of three bodies

is far away it will continue to move in the same elliptical orbit (denoted by *1* in the figure).

But if the initial positions of the meteoroid and Jupiter and their velocities are such that at this instant they make a close approach to each other, the Jovian gravitational pull on the meteoroid will be so strong that the small body will deviate from its initial path. The magnitude and nature of this deviation depend on the mutual positions of the sun, Jupiter and the meteoroid at the instant of closest approach, and on the magnitudes and directions of their velocities, which, in the final analysis, means that everything depends on the initial conditions of the problem. It is even possible that the perturbation produced by Jupiter will lead to such an increase in velocity that the body will recede from the sun along either a hyperbolic or parabolic path. Let us assume meanwhile that the meteoroid's velocity has been altered so that after Jupiter recedes the meteoroid settles in an ellipse denoted by number *2* in Fig. 33. This elliptical orbit is quite different from the first one. In this orbit the meteoroid should return, in equal intervals of time, to the very same spot at which it encountered Jupiter. If the orbital periods of M and J are not equal the second encounter will not occur at once. Nevertheless, sooner or later the meteoroid and Jupiter will meet. And again the orbit of the small body will be altered radically; it will begin to move in an entirely different ellipse or will settle into a hyperbola or parabola.

Slight variations in the conditions of the first encounter may lead to comparatively great differences in the conditions of the second encounter, that is, to a big difference in the size of the new orbit and in the period of revolution. This difference will lead to a still greater alteration of the conditions of the third encounter, which will undoubtedly take place if the new orbit is an ellipse, and so forth. Any general equation that would take into account all the possible consequences of each encounter and their diversity in the case of very slight changes in the initial conditions will be incredibly complicated.

Jupiter will of course alter the orbit of the meteoroid in between encounters as well as during encounters, which is an added complication of the problem. And if M had a considerable mass there would be the added difficulty of M's influence on Jupiter and the associated alteration of the Jovian orbit. By now it is probably clear that the mathematical difficulties that arise in solving the three-body problem are due to the complexity and extraordinary diversity of the motions that can occur, given different initial positions and velocities of M.

It is true that in 1912 a Finnish mathematician, Sundmann, found a theoretical solution of the three-body problem. But the mathematical equations that he obtained are so involved that they do not permit calculating the positions of the bodies in space or drawing any conclusions about the properties or nature of the motions, so that Sundmann's equations are, as yet, of no practical value.

The result is that to this day we are not in possession of any complete solution of the problem of the motion of three or more bodies. Therefore, various approximate methods are used in studying the motions of bodies in the solar system.

11. SUCCESSIVE APPROXIMATIONS IN THE THEORY OF MOTION OF HEAVENLY BODIES

The motions of heavenly bodies have been investigated chiefly by methods of *quantitative* celestial mechanics. These methods enable us to find such an approximate solution to the problem of the motions of certain bodies as is close to the unknown exact solution over a given interval of time. Quantitative celestial mechanics permits of building a *theory of motion* of astronomical bodies. Applying these theories it is possible to compute the positions of bodies in space in the course of a certain period of time, to investigate the nature of the mutual influence of different bodies over a given interval of time, to establish a relationship between the mutual perturbations of the bodies and their masses, and then determine these masses. In short, then, the theory of motion enables one to describe fully the motions of celestial bodies in the course of specific time intervals, starting from an initial instant.

There are different concrete methods of constructing theories of motion. One is the method of successive approximations which we shall apply to the problem of planetary motion.

The most conspicuous feature of planetary motion is that the planets pursue elliptical paths that are very close to unperturbed elliptical motion. This is due to the fact that the solar mass is much greater than that of all the other bodies of the solar system, which means that each planet is more strongly attracted by the sun than by any other body in the system. Let us find out how much more the sun attracts the earth than does Jupiter—the largest of the planets. Jupiter is roughly five times farther from the sun than the earth, and the closest earth-Jupiter distance (at opposition) is some four times the sun-earth distance. Noting that the Jovian mass is about 1 /1000 that of the sun, we find that at closest approach Jupiter pulls the earth with a force that is $4^2 \times \times 1,000 = 16,000$ times weaker than that produced by the sun. Jupiter's attraction is still less at other points in the orbit, and the gravitational pull of the other planets on the earth is much weaker.

In the same way we can find out how much weaker is the attraction of any other planet at closest approach (at opposition) than that of the sun. These figures are given in Table 3 on p. 76. Also listed in the table are up-to-date figures on the planetary masses and the mean acceleration imparted to each planet by the sun.

The numbers in the rows of the table indicate how many times stronger is solar attraction on a given planet than that of the other planets at closest approach. For example, the gravitational pull of the sun on the earth exceeds that of Mercury 3 million times, of Venus 32,000 times, Mars 800,000 times, and Jupiter 16,000 times. The numbers in each column show the influence of the given planet on any other planet (at closest approach). For example, the sun exerts a gravitational pull on Mercury that is 154,000 times that of Jupiter; the corresponding figures for the other planets are: on Venus, 40,000; Earth, 16,000; Mars, 6,500; Saturn, 200; Uranus, 500: and Neptune 700 times.

Thus, the principal force governing the motion of bodies of the solar system is the gravitational attraction of the sun. The effect of mutual planetary attractions is small in comparison to that of the sun's pull. This is the reason why the planets move about the sun almost in ellipses and experience but slight deviations (perturbations) from their elliptical paths.

Actually it was this that enabled Kepler to discover his remarkable laws. If the planetary masses were greater and if they exerted a more disturbing influence on each other their paths would differ so greatly from the elliptical that these laws would not be valid.

The method of successive approximations exploits the fact that the planets pursue almost unperturbed elliptical orbits so that Kepler's laws permit finding the approximate position of a planet in an unperturbed orbit at any instant of time. An approximate knowledge of the planetary configuration enables one to compute the forces of mutual attraction and the resulting acceleration of the planets for each instant of time. It is these additional accelerations, which combine with planetary accelerations produced by the sun, that are the *disturbing accelerations*. They define not the paths themselves of the planets, but deviations from known elliptical paths. These disturbing accelerations may be used to determine the perturbations for each instant of time. These will be what is known as per-

Table 3

	Mercury	Venus	Earth	Mars	Jupiter	Saturn	Uranus	Neptune	Fraction of solar mass	Mean acceleration of planet towards sun
Mercury	—	$300 \cdot 10^3$	$800 \cdot 10^3$	$24 \cdot 10^6$	$154 \cdot 10^3$	$2 \cdot 10^6$	$60 \cdot 10^6$	$100 \cdot 10^6$	$\frac{1}{7,500,000}$	3.7 cm/sec²
Venus	$15 \cdot 10^7$	—	$53 \cdot 10^3$	$9 \cdot 10^6$	$40 \cdot 10^3$	$540 \cdot 10^3$	$17 \cdot 10^6$	$36 \cdot 10^6$	$\frac{1}{408,000}$	1.2
Earth	$3 \cdot 10^6$	$32 \cdot 10^3$	—	$800 \cdot 10^3$	$16 \cdot 10^3$	$250 \cdot 10^3$	$7.5 \cdot 10^6$	$18 \cdot 10^6$	$\frac{1}{333,420}$	0.6
Mars	$4 \cdot 10^6$	$120 \cdot 10^3$	$43 \cdot 10^3$	—	$6,500$	$105 \cdot 10^3$	$3.5 \cdot 10^6$	$14 \cdot 10^6$	$\frac{1}{3,093,500}$	0.26
Jupiter	$6 \cdot 10^6$	$300 \cdot 10^3$	$200 \cdot 10^3$	$1.5 \cdot 10^6$	—	$2,500$	$170 \cdot 10^3$	$460 \cdot 10^3$	$\frac{1}{7,047}$	0.022
Saturn	$7 \cdot 10^6$	$350 \cdot 10^3$	$240 \cdot 10^3$	$2 \cdot 10^6$	200	—	$25 \cdot 10^3$	$88 \cdot 10^3$	$\frac{1}{3,500}$	0.006
Uranus	$7.5 \cdot 10^6$	$400 \cdot 10^3$	$260 \cdot 10^3$	$2.4 \cdot 10^6$	500	900	—	$7 \cdot 10^3$	$\frac{1}{22,870}$	0.0015
Neptune	$7.5 \cdot 10^6$	$400 \cdot 10^3$	$300 \cdot 10^3$	$3 \cdot 10^6$	$70C$	$1,540$	$3 \cdot 10^3$	—	$\frac{1}{19,500}$	0.00067

turbations defined *in the first approximation* or *perturbations of the first order.* Their magnitudes are proportional to the masses of the disturbing bodies.

Perturbations of the first order do not coincide with the full (precise) planetary perturbations since the disturbing accelerations are computed for approximate positions of the planets. Still, insofar as the planetary deviations from these positions are slight, perturbations of the first order will not differ very much from precise perturbations.

An account of first-order perturbations permits computing the new positions of the planet in space (first approximation) for each instant of time. These new, approximate positions will be more accurate than those computed on the basis of the equations of elliptical unperturbed motion, since the deviations from elliptical motion have in large measure been taken into consideration.

Utilizing this new and more accurate planetary configuration for each instant of time, it is again possible to compute the mutual attractions and disturbing accelerations of the planets for these instants, and then also the perturbations. These perturbations determined in what is known as the *second approximation* will now be closer to the precise values than those of the first order. The planetary positions can now be determined more accurately than with account taken of first-order perturbations (second approximation). Similarly, it is possible to calculate the perturbations in the third approximation, etc.

The smaller the mass of the disturbing bodies—planets—as compared to that of the principal attracting body (the sun), the fewer approximations are required in order to obtain a more or less precise result. Let us consider a planet disturbed by another planet. In the first approximation, the disturbing acceleration of our planet at each instant will be computed from the equation

$$w_1 = \frac{m}{r_o^2}$$

where m is the mass of the disturbing planet (in a fraction of the solar mass), and r_o is the distance between the two planets calculated on the condition that the planets pursue unperturbed elliptical paths. These disturbing accelerations are used to compute the perturbations in the first approximation, which, in magnitude, will be proportional to the disturbing mass m. With these accounted for we obtain more accurate values of the distances between the planets at each instant of time. We denote these distances by r_1. The r_1 values will differ from the corresponding r_0 values by amounts proportional to the mass m:

$$r_1 = r_0 + \Delta r_1; \quad \Delta r_1 \sim m.$$

In the second approximation, the disturbing acceleration is computed from the equation

$$w_2 = \frac{m}{r_1^2}.$$

What is the difference between accelerations w_1 and w_2 at one and the same instant of time? To find out, we take the difference

$$w_1 - w_2 = \frac{m}{r_0^2} - \frac{m}{(r_0 + \Delta r_1)^2} = m \, \frac{2\Delta r_1 + \Delta r^2_1}{r_0(r_0 + \Delta r_1)^2} = m \, \Delta r_1$$
$$\frac{2 + \Delta r_1}{r_0(r_0 + \Delta r_1)^2}$$

Since $\Delta r_1 \sim m$, the difference $w_1 - w_2$ is now proportional to the *square* of the disturbing mass m. Using accelerations w_2 to compute the perturbations in the second approximation we obtain new distances at each instant $— r_2$. They will differ from r_1 by magnitudes that are proportional to the square of the disturbing mass m, that is,

$$r_2 = r_1 + \Delta r_2, \quad \Delta r_2 \sim m^2.$$

We now calculate the disturbing accelarations in the third approximation:

$$w_3 = \frac{m}{r_2^2}$$

and write the difference

$$w_2 - w_3 = \frac{m}{r^2_1} - \frac{m}{(r_1 + \Delta r_2)^2} = m \, \Delta r_2 \frac{2 + \Delta r_2}{r_1(r_1 + \Delta r_2)}.$$

We see that this difference is proportional to the *cube* of the perturbing mass m. The mutual distances computed in the third approximation (r_3) will differ from r_3 by magnitudes that are proportional to m^3, that is,

$$r_3 = r_2 + \Delta r_3, \quad \Delta r_3 \sim m^3.$$

Summing up, then, we see that the first approximation enables us to determine the position of a planet with an accuracy to magnitudes proportional to m, the second approximation, to m^2, the third approximation, to m^3, etc. The smaller m is, the greater the accuracy of the successive approximations, the less the positions of the planet in the first and second approximations differ from the precise positions of the planet in the given problem.

As we have already noticed, the masses of the planets are comparatively small. Jupiter has the largest mass, and yet this is only about 1/000 the solar mass. This is why the method of successive approximations is very convenient in studying planetary motion. Let us carry out some numerical calculations for the three-body problem: Sun-Jupiter-Saturn. We shall assume that neither Jupiter nor Saturn are influenced by the other planets.

Saturn and Jupiter revolve round the sun in ellipses that are close to circles with radii of 1,500,000,000 km. and 750,000,000 km. respectively. Jupiter completes a circuit in 12 years, Saturn in 30 years. First, we evaluate the perturbations that Saturn experiences due to Jovian gravitation and that Jupiter experiences due to the attraction of Saturn in the course of three years from the time of opposition. To do this, we calculate the disturbing accelarations that these planets impart to each other at the time of opposition. According to Table 3 (p. 76), at the times of opposition solar gravitational action on Saturn is roughly 200 times stronger than that of Jupiter. Since acceleration is proportional to the force of attraction, Jupiter, at this time, imparts to Saturn an acceleration 1/200 that of the sun. The table gives the mean acceleration of Saturn due to the sun's attraction: 0.006 cm/sec². Thus, the disturbing acceleration of Saturn produced by Jupiter at opposition is approximately

$$w_{\text{J}} = \frac{0.006}{200} = 0.00003 \text{ cm/sec}^2.$$

In exactly the same way we calculate the disturbing acceleration of Jupiter due to Saturn, w_{S}, at opposition, noting that Saturn attracts Jupiter with a force roughly $^1/_{2,500}$ that of the sun, while the mean Jovian acceleration due to the sun's gravitational pull is 0.022 cm/sec². We then find

$$w_{\text{S}} = \frac{0.022}{2,500} = 0.000009 \text{ cm/sec}^2.$$

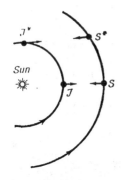

Fig. 34. The mutual perturbations of Jupiter and Saturn

During three years, Jupiter covers about 1/4 of its orbital path, while Saturn moves 1/10 of its way around the sun (Fig. 34). Three years from opposition these planets will occupy positions J^* and S^* in Fig. 34. Beginning with opposition, the distance between Jupiter and Saturn will continually increase, and the disturbing accelerations will consequently diminish. But the directions of the disturbing accelerations will not change much during these three years, and, without committing a gross error, we can take the disturbing accelerations during this time to be constant and equal to their maximum values—0.00003 cm/sec² and 0.000009 cm/sec², respectively—and in the same direction. Then the deviations from elliptical paths calculated by the well-known equation for uniformly accelerated motion

$$s = {}^1/_2 \, wt^2$$

would, in any case, be greater than the true values. Computing these deviations in kilometres, we obtain for Saturn

$$s \approx 15 \times 10^{10} \text{ cm.} = 1.5 \text{ million km.}$$

and for Jupiter

$$s \approx 5 \times 10^{10} \text{ cm.} = 500,000 \text{ km.}$$

Consequently, in the course of three years Saturn does not deviate from its elliptical path more than 1,500,000 km., and Jupiter not more than 500,000 km. Of course these distances are not small, but in comparison with the solar distances from these planets they are rather insignificant. As viewed from the earth, these deviations of Jupiter and Saturn in space correspond to deviations in their apparent positions in the sky that do not exceed 3' for Saturn and 2' for Jupiter.

To obtain the first approximation, a precise calculation is made of the disturbing accelerations of Jupiter and Saturn and then of their perturbations at each instant of time during the given three years on the condition that these planets are moving in true ellipses. But we shall do otherwise and more simply.

Since the total perturbations of Saturn and Jupiter during three years do not exceed 1,500,000 and 500,000 km. respectively, we can estimate to what extent the first-approximation perturbations differ from the full perturbations. Indeed, the precise positions of Jupiter and Saturn can differ from the positions in elliptical orbits, for which we computed the disturbing accelerations in the first approximation, by not more than 1.5 million km. and 0.5 million km., respectively, while the true distances between them cannot differ by more than $1.5+0.5=2$ million km. from those taken in the calculations. The accelerations due to the mutual gravitational action of the planets vary as the square of the distances. Therefore,

$$\frac{w}{w_1} = \frac{r^2{}_1}{r^2}$$

where w_1 and r_1 are the disturbing acceleration and the distance between the planets in the first approximation, while w and r are the true disturbing acceleration and distance between Jupiter and Saturn.

Applying the rule of proportions, we write the following relation:

$$\frac{w - w_1}{w_1} = \frac{r^2{}_1 - r^2}{r^2}$$

Since r_1 cannot differ from r by more than 2 million km. and r by not less than $1,500-750=750$ million km., this ratio does not exceed 1/180. Therefore, w_1 does not differ from w by more than one part in 180. Otherwise stated, in the first approximation, the error in disturbing accelerations does not exceed 1/180. And for this reason, the first-approximation errors of perturbations calculated on the basis of these accelerations do not exceed

$$\frac{15 \times 10^5}{180} = 8{,}300 \ \text{km. and} \ \frac{5 \times 10^5}{180} = 2{,}800 \ \text{km.}$$

respectively.

Thus, account of first-approximation perturbations enables one to determine the positions of Jupiter and Saturn in the course of the given three years with an accuracy to within 8,300 km. and 2,800 km. respectively. Such discrepancies in the motion of Jupiter and Saturn correspond to apparent displacements in the sky, as seen from the earth, that do not exceed 1″. This accuracy in determining planetary positions is quite sufficient, and there is no need to compute perturbations in the second approximation. But if we wished to determine these perturbations we would have to compute again the positions of the two planets obtained after taking into account the perturbations of the first order, then the disturbing accelerations, and finally the perturbations themselves in the second approximation for each instant of time. Since errors in the distances between Jupiter and Saturn will not exceed $2,800 + 8,300 = 11,100$ km., the errors of disturbing accelerations and the errors of perturbation of the second approximation will not exceed one part in 35,000 of these figures. In this case, the errors in the positions of Saturn and Jupiter, after taking into account the second-approximation perturbations, will not exceed

$$\frac{15 \times 10^5}{35,000} \approx 43 \text{ km. and } \frac{5 \times 10^5}{35,000} \approx 14 \text{ km.}$$

respectively. An accuracy of this degree is more than sufficient for all practical applications of the theory.

On what will the expressions for perturbations depend? The positions of Jupiter and Saturn in space in elliptical motion (this type of motion was one of the starting conditions) are determined by the elements of their elliptical orbits and by the time; the mutual disturbing accelerations are proportional to the masses of the planets. Therefore, the computed perturbations of the first, second, etc., approximations will depend solely on the initial orbital elements, the time and the masses of Jupiter and Saturn. The equations of our theory of motion will relate the positions in space of Jupiter and Saturn for each instant of time, the initial orbital elements and their masses, and in this way we will be able to construct so-called *analytical theories* of Jupiter and Saturn.

Such, essentially, is the method of successive approximations which has found wide and successful application in the construction of theories of motion of planets and other bodies of the solar system. This was the procedure used to build the analytical theories of motion of all the major planets.

Analytical theories of motion are valuable not only because they permit of calculating the apparent positions of heavenly bodies, but because they enable one to study the character of the mutual influences of planets and other bodies and also to calculate their masses.

We have already learned that the masses of planets that have satellites can be determined approximately by means of Kepler's Third Law. This was the procedure used to find the masses of the earth, Mars, Jupiter, Saturn, Uranus, and Neptune. Now how is one to find the mass of a planet not attended by satellites? By means of the analytical theories of these planets.

By way of illustration, let us determine the mass of Venus. We may make an approximate estimation of the Venusian mass on the assumption that its mean density is the same as that of the earth. The size of the planet is known, so we can calculate the approximate value of its mass.

This figure will, of course, be very inaccurate. Refinements in this value can be made by considering the perturbations that Venus produces in the motion of the earth.

We already know that the mutual perturbations of bodies depend upon their masses. Let us calculate the perturbations that Venusian gravitation produces in the earth's motion using Venus' mass approximated by the procedure described above. We then compare the calculated perturbations with those actually observed in the earth's motion (perturbations caused by the other planets should naturally have already been taken into account).

Since the mass of the planet has not been accurately determined, the computed perturbations will differ from the observed ones. By varying the mass of Venus it is possible to select a value for which the computed perturbations will least of all differ from the observed ones. The result will be a Venusian mass closest to reality.

It is far more difficult to determine the mass of Mercury and Pluto. Both of these planets are very small and produce only minute perturbations in the motions of the other planets. This is the reason why the masses of these two planets are so imperfectly known at the present time.

12. THE DISCOVERY OF NEPTUNE

One of the most brilliant attainments of celestial mechanics was the discovery of the planet Neptune.

Five planets—Mercury, Venus, Mars, Jupiter and Saturn—had been known since remotest antiquity.

In 1781 the English astronomer W. Herschel, while scanning the heavens with his telescope, noted a faint star that moved slowly among the fixed stars. Herschel thought it to be a comet. However, calculations carried out by the Russian astronomer A. I. Leksel showed that the new body was moving about the sun almost in a circle with the solar distance roughly twice that of Saturn. This was a new major planet of the solar system; it was called Uranus.

Theoretical studies of the planet's motions based on the Newtonian law of gravitation came up against unexpected difficulties. Uranus refused to obey strictly the law of gravitation. Theories of its motion, with account taken of the perturbations produced by all the known planets, were not capable of representing the observed motion accurately. The calculated positions of the planet deviated regularly from the apparent positions. In 1830 these deviations amounted to about 20″, in 1840 to 1′.5, and in 1844 to about 2′. By this time Newton's law had so firmly "entrenched" itself that only a few astronomers ascribed these deviations to a breakdown in the law. Other causes were sought, and it was suggested that beyond the orbit of Uranus was another planet which produced these additional perturbations in Uranus' orbit. On this conjecture, the discrepancies between theory and observation were explained by disturbances of Uranus caused by this unknown planet.

Naturally, to locate this conjectured planet by telescope was very difficult. It could only be hit upon by accident. This brought up the problem of determining the motion of the planet, that is, its orbit, from the perturbations that it was thought to produce in the motion of Uranus. This problem could be solved by means of an analytical theory of the motion of Uranus with account taken not only of all the disturbing attractions of all the known planets but also the disturbing influence of the unknown planet. Above we pointed out that in constructing an analytical theory of the motion of a planet one can obtain the relationship between the perturbations of the given planet and the initial orbital elements or, in other words, the approximated paths of those planets that disturb the given planet by their gravitational attraction. Up till now the problem was to find the perturbations from known approximate paths of the planets. In contrast, the new problem was to find the orbital elements of an unknown disturbing planet by means of equations of the analytical theory of Uranus and on the basis of the known perturbations of Uranus.

This work, which involved tremendous mathematical difficulties due to the complexity of the analytical theories of the planets and the smallness of Uranus' perturbations caused by the unknown planet, was carried out almost at the same time by the English astronomer Adams (in 1843-45) and the French astronomer Leverrier (in 1845-46). After the orbital elements of the planet had been found it was possible to derive its apparent path in the sky and its positions at any instant of time. On September 23, 1846, Galle, of Berlin Observatory, aimed his telescope at that part of the sky where, according to Leverrier's data, the unknown planet should be. At a distance of about only 1° from the spot predicted by Leverrier, Galle actually detected a new body with a small planetary disk. A few days later it was found that this body was in motion among the stars. This was the new planet that later received the name of Neptune.

This "paper and pencil" discovery of Neptune was a new and very convincing proof of the correctness of Newton's law.

13. PERIODIC AND SECULAR PERTURBATIONS

An analysis of the theories of planetary motion confirmed by observational material permits us to distinguish between two types of perturbations: *periodic* and *secular*. The former are periodically repeating oscillations about the elliptical motion (the orbital element oscillates about a certain mean value). The latter are distinguished by progressively mounting deviations from unperturbed motion (the magnitude of the element continually diminishes or increases).

The origin of the periodic perturbations may be explained

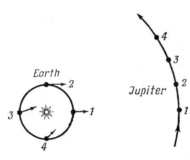

Fig. 35. An explanation of periodic perturbations of the earth with a period of one year

by noting that the planetary configurations, and consequently the direction and magnitude of the disturbing accelerations vary rather rapidly and periodically, with the result that one planet accelerates another first in one direction and then in the other.

To illustrate, let us see how the disturbing accelerations of the earth produced by Jupiter vary in the course of one year. During this time, the earth completes a single circuit while Jupiter does only 1/12 of one revolution about the sun. The arrows in Fig. 35 indicate the directions of terrestrial disturbing accelerations due to Jupiter each quarter year. It will be clear from the drawing that at position 2 the earth's orbital velocity is reduced due to Jovian gravitation, while in position 4 (a half year later) Jupiter's attraction should increase the velocity of the earth's motion. The result is periodic perturbations of the earth with a period of one year. Besides these periodic annual perturbations there will be others too. For instance, Fig. 35 shows that during the year Jupiter's attraction displaces the earth towards Jupiter. Let us now see how this direction of annual terrestrial displacement will vary during one Jovian orbital period equal to 12 years.

In Fig. 36 the arrows indicate the direction of the annual displacement of the earth due to Jupiter for each year of the 12-year period. We see that in 12 years our arrows move 360°, which means that each annual displacement in one direction corresponds (six years later) to a displacement in the opposite direction. These perturbations have periods of 12 years. We have thus discovered two components of the earth's deviations from unperturbed motion due to the attraction of Jupiter: periodic oscillations of one and twelve years.

As a rule, periodic perturbations of the planets are small and do not lead to considerable alterations in their motions. The maximum apparent deviations in the sky due to periodic perturbations from positions that correspond to elliptical motion are: for Mercury about 15″, for Venus 30″, for the earth 1′, Mars 2′, Uranus 3′, and for Neptune 1′.5. Only for Jupiter and Saturn do these deviations attain a considerable magnitude—28′ and 48′ respectively. The period of these perturbations is also great, roughly 900 years.

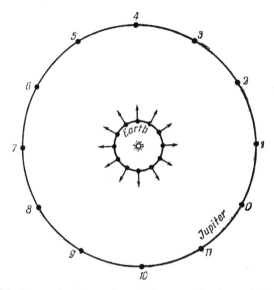

Fig. 36. An explanation of periodic perturbations of the earth with a period of 12 years

Perturbations with such long periods are called *long-period* perturbations. They occupy a borderline position between the periodic group and secular perturbations. It often happens that we are unable to distinguish them directly in observations, since if the observations embrace a period that is much less than that of the long-period perturbations, the latter, just like the secular type, will be in the nature of continually increasing deviations. For example, a very slow variation in the mean orbital velocities of motion of Jupiter and Saturn was noted in the seventeenth century, but since the period of these variations is close to 900 years, observations alone are not enough to decide whether the perturbations are periodic or secular. Their periodicity is established by theory, and observations confirm the fact that these perturbations do actually occur as predicted by the theory of motion. What produces the long-period perturbations of these planets?

The orbital periods of Jupiter and Saturn about the sun are known to be approximately 12 and 30 years so that the ratio of these periods is close to 2/5: Saturn completes two circuits during roughly the same time that Jupiter does five. Such orbital periods are said to be *commensurable*.

Commensurability accounts for a phenomenon similar to what in mechanics is called *resonance*. Resonance occurs when an additional periodic force acts on an oscillating body in rhythm with the latter's oscillations. Even if this force is very small it can gradually build up a considerable amplitude.

Saturn's motion round the sun can be regarded as periodic oscillations about the sun with a period of 30 years. After every two circuits that Saturn completes, the latter, Jupiter and the sun appear in the same configuration as 30 years before, so that the perturbing action of Jupiter on Saturn is repeated regularly. In other words, the perturbing action of Jupiter on Saturn is periodic with a period twice Saturn's orbital period.

To summarize, Saturn is in periodic oscillation relative to the sun, and is acted upon in resonance by a periodic perturbing force (perturbations caused by the planet Jupiter). The same may be said of the disturbing influence of Saturn on Jupiter. The result is a situation which leads to resonance. This is why the mutual perturbations of Jupiter and Saturn associated with the commensurability of their periods are so great.

Besides periodic, we also find secular perturbations. These force the planet farther and farther away from its unperturbed course. Analytical theories of planetary motion contain such perturbations that vary in proportion to the time and even to the square and cube of the time. However, these perturbations increase very slowly. For example, according to the theories of motion of the earth, Venus, and Jupiter constructed by Leverrier, the eccentricities of the orbits of these planets have secular perturbations expressed by the following formulae:

$e = 0.0167498 - 0.0000426t - 0.000000137t^2$ (Earth)
$e = 0.00681636 - 0.00005384t + 0.000000126t^2$ (Venus)
$e = 0.04833475 + 0.000164180t - 0.000000468t^2$ (Jupiter)

In these formulae, t is the time reckoned in centuries from 1900. It may be seen that during one year the eccentricities of these planets vary by only 0.0000004; 0.00000054; 0.000001641 respectively, and in 100 years by 0.0000427; 0.0000537; 0.00016371.

However, such perturbations acting over long periods of time could lead to essential changes in the planetary motions. If, for example, we take a period of 10 000 years, the eccentricity of the earth's orbit should diminish to 0.0134, of the orbit of Venus—to 0.0014, while the eccentricity of the Jovian orbit should increase to 0.064.

But we should not lose sight of the fact that the analytical theories of planetary motion are capable of representing the actual motions of the planets with sufficient accuracy for only a definite interval of time, beyond which these theories do not hold. Unfortunately, in only a few special cases is it possible to state theoretically the interval of time during which discrepancies between theory and the precise solution of the problem of motion of the given bodies are still sufficiently small.

Ordinarily, the developed theory of motion of some body is compared with all available observations. This naturally yields divergences between the theory and the observations. The magnitude of these divergences and the interval of time embraced by the comparison permits one to judge of errors in the theory.

Comparisons of modern analytical theories of planetary motion with observational data covering the time period approximately from 1800 to the present day exhibit departures from observations that amount to within several seconds of arc. Thus, we may presume that the present theories of planetary motion will hold for calculations during 100 200 years and will allow us to predict the positions of the planets in the sky during this period with an accuracy to several seconds of arc. During this time, the orbital elements of the planets will vary in accordance with the equations of these theories, but it is not known whether they will continue to vary in the same way in the future. At any rate, there can be no question of using these equations to study planetary motion in the course of many thousands of years. To illustrate, if we calculated the eccentricity of Venus' orbit for 20,000 years hence using the above equation, it would be negative. But this is impossible since the eccentricity of an orbit is a positive quantity or zero. This equation is therefore useless for calculations of the Venusian eccentricity tens of thousands of years in advance.

The conclusion is that analytical theories of planetary motion are not suited for studies of purely secular perturbations. But celestial mechanics has worked out other methods, designed not for calculating the precise positions of heavenly bodies but specially for the investigation of secular perturbations. Application of these methods has shown that some of the perturbations, which in the analytical theories of planetary motion are regarded as secular, are actually of the long-period type. Such are variations of eccentricities and inclinations of planetary orbits. Mutual attractions of the planets periodically alter the shape of the orbits and their inclinations to the plane of the ecliptic. It appears that these perturbations of the eccentricities and inclinations represent a set of oscillations with different and very long periods, of the order of tens of thousands of years. Such perturbations are not regularly periodic but rather in the nature of very, very slow irregular oscillations.

Thus, the theory of secular perturbations has shown that the variations of the eccentricities and inclinations of the planetary orbits are not exactly secular. But since the periods of these variations attain scores and hundreds of thousands of years, the literature has retained the term "secular."

As an illustration, Table 4 lists the secular variations (for 100,000 years before and after 1850) of the mean values of two elements of the variational orbits of Mars and the earth: the eccentricity and inclination calculated by Leverrier (neglecting short-period perturbations). The orbital inclinations are reckoned relative to the plane of the earth's motion in 1850.

The trend in the variation of the elements is graphically displayed in Figs. 37 and 38, which are based on Table 4.

Table 4

t, thousands of years	Mars		Earth	
	e	*i*	*e*	*i*
−100	0.1079	3°13′45″	0.0473	3°45′31″
90	0.1195	2 55 36	0.0452	2 42 19
80	0.1251	1 55 12	0.0398	1 18 58
70	0.1225	30 01	0.0316	1 13 58
60	0.1175	1 01 41	0.0218	2 36 42
50	0.0978	2 09 32	0.0131	3 40 11
40	0.0832	2 46 15	0.0109	4 03 01
30	0.0746	2 54 43	0.0151	3 41 51
20	0.0840	2 46 37	0.0188	2 44 12
−10	0.0884	2 27 51	0.0187	1 24 35
0*)	0.0932	1 51 06	0.0168	0 00 00
+10	0.1006	49 17	0.0115	1 14 26
20	0.1036	53 49	0.0047	2 07 46
30	0.1013	2 29 09	0.0059	2 33 19
40	0.0945	3 49 17	0.0124	2 27 53
50	0.0857	4 27 27	0.0173	1 51 54
60	0.0797	4 10 49	0.0199	51 52
70	0.0825	3 05 11	0.0211	34 35
80	0.0948	1 46 11	0.0188	1 45 40
90	0.1113	1 55 26	0.0176	2 40 56
+100	0.1258	49 45	0.0189	3 02 57

From the table and graphs it follows that the eccentricity of the earth's orbit fluctuates between zero (circular orbit) and 0.069, while the inclination of this orbit can increase to 4°41′.

Leverrier computed such tables for other planets too. Table 5 gives the maximum and minimum values of the eccentricities and also the maximum values of the orbital inclinations of the major planets relative to the orbital plane of the earth in 1850.

These "secular" perturbations of the eccentricities and orbital inclinations of the planets should lead to rather noticeable changes in the apparent positions of the planets in the sky. For Venus, these changes (due to increasing orbital eccentricity to $e=0.071$) will reach 7 8°, while for Mars (due to an increase in the inclination) they will be at least 5°.

---

* The initial instant of time $t=0$ is taken at 1850.

Still, these perturbations are not great enough to alter considerably the type of motion of the planets. Their orbits will still be close to circles and the orbital planes will, as before, be only slightly inclined to each other.

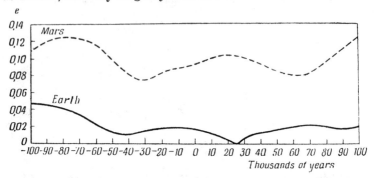

Fig. 37. Variations in the eccentricities of the orbits of Earth and Mars during 200,000 years

Alterations of a more radical nature in the planetary motions could, in time, result from the purely secular changes of the eccentricities, inclinations, and particularly the semi-major axes of the orbits. A secular increase in the semi-major axes would, for example, mean that all the planets were gradually receding from the sun.

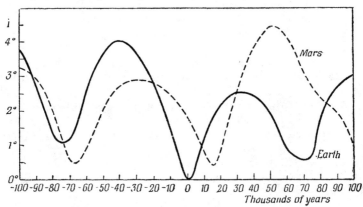

Fig. 38. Variations in the orbital inclinations of Earth and Mars during 200,000 years (relative to the earth's orbital plane in 1850)

Table 5

	e		*i* maximum
	minimum	maximum	
Mercury	0.121	0.232	9°17'
Venus	0	0.071	5°18'
Earth	0	0.069	4°41'
Mars	0.018	0.140	7°09'
Jupiter	0.025	0.061	2°01'
Saturn	0.012	0.084	2°33'
Uranus	0.012	0.078	2°33'

However, neither theoretical investigations nor observations have yet detected true secular changes in these orbital elements. Of course, the semi-major axes of the planetary orbits, and also the eccentricities and inclinations do not remain the same. They experience slight periodic fluctuations; however, in the course of many hundreds of thousands of years there should not be any constant decrease or increase in the semi-major axes. In other words, the planets will be moving just about as they are now for a very, very long time.

Of all the orbital elements of the planets only two—the longitude of the ascending node and the perihelion distance from the node—are subject to secular variations. The ascending nodes of all the planets "retreat," that is to say, they move in a direction opposite to that of the orbital motion of the planets. On the other hand, the perihelia of the majority of planets move in the same direction as the planets themselves. But these motions are very slow. "Fastest" is the line of apsides of Saturn's orbit. It completes a full circuit in 57,000 years. Jupiter's perihelion makes a complete circuit 1/6 as fast—in 349,000 years.

14. NUMERICAL METHODS IN CELESTIAL MECHANICS

In addition to analytical methods, wide use in the study of the motions of planets and other heavenly bodies is being made at present of *numerical* methods. They differ from the analytical approach in that with their aid one obtains not equations that define the perturbations of bodies as a function of time, but only numbers (coordinates) that indicate the position of a body in space at definite instants of time. How are these specific positions found?

Let several bodies attract each other by Newton's law. If for each body we know the position and velocity at the initial instant of time t_o, it is possible to determine the forces with which these bodies act on one another and the accelerations that they impart to each other at the initial instant. We select an instant t_1 that is close to the initial instant and assume that during the small interval of time $\Delta t = t_1 - t_o$ the accelerations of the bodies do not change. Then, applying the equations of uniformly accelerated motion we can calculate for each body its departure from uniform and rectilinear motion during the time Δt and its position and velocity at the instant t_1. From the new positions of the bodies it is again possible to calculate the forces acting between them and the accelerations at the instant t_1; it is then possible once more to determine their positions and velocities at the subsequent close instant t_2, etc.

It is thus possible by successive steps to calculate the approximate positions of heavenly bodies in space and to compile a table that indicates the positions of these bodies at instants t_1, t_2, t_3, . . . during a specific interval of time. This is how a *numerical theory* of the motion of bodies is constructed for this interval of time.

These computations of the successive positions of a heavenly body in its orbit are rather simple, requiring only the four ordinary operations of arithmetic. However, the number of these operations that must be carried out in order to derive the motion over a more or less appreciable interval of time (for example, 10 to 20 years) is stupendous. This is why numerical techniques have become widespread only during the past ten to fifteen years, since the advent of modern calculating machines.

In 1951, tables were published in the United States of the motions of the four largest planets of the solar system—Jupiter, Saturn, Uranus, and Neptune. These tables, compiled by means of electronic computers, indicate the positions of the planets from 1653 to 2060 for every 40 days. A comparison of the observations of these planets from 1780 to 1940 with the numerical theory of their motions shows that this theory accords better with observation than do the most refined analytical theories created by the American

94

astronomers Hill and Newcomb. The numerical theory enables one to represent the apparent motions of these planets with the same accuracy as the planetary positions determined by observations. On the average, divergences between numerical planetary theory and observational data do not exceed one second of arc.

Numerical methods are now being successfully applied in studies of the motion of asteroids and satellites as well as in solving other problems of celestial mechanics. As a rule, they enable us to predict, with far greater accuracy than the analytical methods, the apparent positions of celestial bodies. And what is more, the most involved problems reduce to simple calculations; no fundamental mathematical difficulties arise. Problems of three and four bodies are solved numerically with the same ease as the two-body problem—only the number of operations increases. This is the great merit of numerical techniques. What demerits have they?

Like analytical methods, the numerical theories of motion cover only a specific and small interval of time. True, calculations may be carried out for any interval of time, no matter how large. But, not only is it, practically speaking, too difficult to create a numerical theory of motion of a body, say, for 100,000 years in advance, but there is also the question of errors in the theory. This is natural since we obtain only an approximate solution of the problem of motion. And to determine its errors for points of time that are far removed from the initial instant (for instance, many thousands of years), or to what extent the position of a given body at this instant (as indicated by the numerical solution) differs from the position given by an exact solution of this problem of motion, is very difficult and, in many cases, even impossible as yet.

Even to begin computations we need to know precisely the masses of the bodies under consideration and we must have complete data on the gravitation of these bodies. Another thing is that in numerical calculations we obtain directly the *perturbations* of these bodies, but not the relationship between these perturbations and quantities that describe the masses, orbital elements and other properties of the perturbing bodies and their motions, as is the case when applying analytical methods. If we did not have at our disposal theoretical methods of motion, we might to this day be ignorant of the mass of Venus or Mercury, and to discover Neptune "theoretically" would be quite out of the question. In themselves numerical methods do not permit of studying the general properties of motion of heavenly bodies.

In conclusion it may be said that numerical methods and numerical theories of motion, as they exist today, are a long way from being all-powerful and cannot replace completely analytical methods and analytical theories of the motion of bodies. Nevertheless, at the present time their practical value is great. And their role and significance will grow in step with future refinements in this field and with improved calculating machines. Papers have already appeared that combine numerical and analytical techniques. Undoubtedly, considerable progress is to be expected in celestial mechanics through the joint application of numerical and analytical methods.

15. SATELLITE THEORY

The satellites move about their primaries in much the same way as the planets do around the sun. Basically, the motion of the satellites of a given planet is governed by the force of gravitation of this planet. If the planet has several attendants (like Jupiter or Saturn) their mutual perturbations have to be taken into account. These perturbations are usually small since the masses of the satellites are small in comparison to that of the planet. The motions of the satellites are also disturbed by the other bodies of the solar system.

Let us compare the perturbing action of the sun and Jupiter on the moons of Saturn.

At closest approach, Jupiter is roughly twice as close to Saturn as the sun. But the mass of the sun is 1,000 times that of Jupiter, and so it attracts a Saturn satellite with $1,000/2^2 = 250$ times the force of Jupiter. The perturbing action of the other planets on Saturn's satellites is still less. For this reason, in satellite theory we ordinarily take into account only the perturbing influence of the sun and disregard the attraction of the planets. Another point is that since the satellites are usually not far from their primaries we must take account of the fact that the planets are not exactly spherical in shape, and that, for this reason, the gravitational pull of the planet does not vary exactly with the inverse-square law.

A study of the mutual perturbations of the satellites enables us to determine their masses. This was how determinations were made of the masses of four of Jupiter's largest satellites and four of Saturn's. The mutual influence of the other satellites of Saturn and Jupiter as well as those of Uranus and Mars was not detected. Their masses can be judged only by their sizes and probable density.

Fig. 39. Saturn

The theories of satellite motion permit of separating out the influence of departures of the planet from sphericity. This influence depends on the compression of the planet, so that observed perturbations enable us to evaluate this compression (so-called *dynamic compression*). However, the motion of satellites is not determined solely by the geometric shape of the planet, but also by the law of the distribution of density of planetary matter, a thing that we do not know. This is why estimates of the dynamic compression of planets are not completely reliable.

Saturn's system of moons is unlike other systems due to the ring. Saturn's ring consists of a multitude of miniature satellites moving approximately in a single plane and so close to one another that at a distance the impression is a solid thin ring.

In building a theory of the motion of Saturn's satellites one has to take into consideration the attraction of this ring. The observed perturbations of the satellites have helped us to determine the mass of the ring, which came out to 1/27,000 the mass of Saturn.

The problem of the motion of the earth satellite—the moon—is one of the most complicated in celestial mechanics. The reasons for this are the following:

1. The moon is the closest heavenly body to the earth. The slightest irregularities in lunar motion are detectable. A shift in the moon's position in space of only 2 km. will result in a 1″ change in its apparent position in the sky. This requires that the accuracy with which the theory predicts lunar positions should be at least as good as this. Theory must be able to define the position of the moon in space to within 2 km. or even more accurately*.

The positions in space of the planets and asteroids may be determined with much less accuracy. For instance, in order to predict the apparent position of Jupiter in the sky to within 1″ it is enough to know its position in space with an accuracy of only 3,000 km.

2. The earth and moon are rather close to the sun and perturbations of the moon due to the sun are very great. Let us calculate how much weaker is the perturbing attraction of the sun than that of the earth, which governs the motion of the moon.

We must make it clear here that the perturbing attraction of the sun is not the force with which the sun attracts the moon. The moon moves round the earth and for this reason the total force with which the sun attracts the moon and earth does not produce any change in the mutual positions of these bodies. A change in the lunar position relative to the earth is due: 1) to the gravitational attraction of the earth (the principal force), and 2) to the fact that the sun's pull on the moon is weaker or stronger than on the earth, that is, to the difference in the force of gravitation exerted by the sun on the earth and on the moon. The ratio of this perturbing difference of forces to the principal force—the earth's pull on the moon—at New Moon, when the latter is closest to the sun, comes out at 1/89, which means that the perturbing attraction is on occasion not so very much less than the principal force. For other satellites, this ratio of principal to perturbing force is many times less.

The earth communicates to the moon an acceleration of about 0.27 cm/sec². The perturbing acceleration of the moon due to the perturbing attraction of the sun is less

---

* To predict the onset of a solar eclipse with an accuracy of 1 sec. it is necessary to know the position of the moon in the sky to within 0″.5.

by a factor of 89 and equal to $0.27/89 \approx 0.003$ cm/sec². Applying equation $s = \frac{1}{2} at^2$ we find that in only three days (during which this disturbing acceleration varies but slightly) the moon will depart 1,000 km. from its unperturbed path, and its apparent position will change by about 4'. And this only in three days! Recall that the perturbations of Saturn and Jupiter do not exceed 3' in 3 years.

3. The earth is not an exact sphere but has the shape of an oblate spheroid. Using the equation of the attraction of a spheroid it is possible to calculate that the perturbing force due to compression does not exceed the earth's gravitational pull on the moon by one part in one million. However, in lunar theory this force has to be reckoned with too.

Thus, on the one hand, the moon experiences relatively strong perturbations of various origin (perturbations due to the sun and planets and also due to the earth's compression); on the other hand, these perturbations have to be calculated with a high degree of accuracy, much higher than is required for those of other celestial bodies. This is what makes the problem of lunar motion so complicated.

Many astronomers and mathematicians, beginning with Newton, d'Alembert and Euler, have constructed theories of lunar motion on the basis of Newton's law of gravitation. In use at the present time is the theory of the moon's motion developed in 1895 by the American astronomer E. W. Brown. This theory permits computing the moon's position in the sky with an error not exceeding $0''.5-1''$. To give the reader some idea of the complexity of lunar theory and of the thoroughness with which it has to be worked out, we may note that some of the equations defining the position of the moon in the sky consist of sums of many hundreds of terms. These terms are periodic perturbations. Most of them do not exceed $0''.1$ and correspond to a 200-metre change in the position of the moon in space.

16. ARTIFICIAL EARTH SATELLITES AND THEIR MOTION

Just recently celestial mechanics was confronted with the motion of a special type of body—artificial earth satellites. On October 4, 1957 the U. S. S. R. launched the first artificial satellite of the earth weighing 83.6 kilograms. The second Soviet satellite (weight—500 kg.) was put into orbit on November 3, 1957, and on May 15, 1958, a third Soviet satellite weighing 1,327 kilograms was launched. Since February 1958 the Americans have put into orbit several small artificial satellites ranging from 1.5 kg. to 67.5 kg. in weight.

After an artificial satellite is launched into orbit about the earth it moves under the influence of the earth's gravitational force just as the moon does. In the first approximation it moves in an ellipse that retains one and the same position in space and at one focus of which is the centre of the earth. Deviations of the satellite from this elliptical (so-called unperturbed) motion—perturbations—are caused primarily by an additional force, the resistance of the terrestrial atmosphere. Besides, the earth does not attract towards its centre in strict accord with Newton's low as a result of its compression and the uneven distribution of the density of material in the earth's interior.

The properties of unperturbed elliptical motion are taken into account primarily during the launching of the satellites. A preliminary calculation is made of the satellite's orbit in space, of its closest and farthest distance from the earth's surface. The launching rocket first moves vertically upwards, then, by means of control systems, it gradually turns in the vertical plane. At a definite, predetermined height the rocket begins to move almost horizontally, and the satellite is detached with a prescribed speed. The velocity of the satellite at this instant determines the shape of its orbit. The plane of the rocket's trajectory will be the orbital plane of the satellite.

If a satellite, at a distance r_0 from the centre of the earth, is given a horizontal velocity exactly equal to

$$V_0 = \sqrt{\frac{fm}{r_0}}$$

where f is the constant of gravitation and m the mass of the earth, it will move in a circle with the centre at the centre of the earth and with a radius of r_0 (see Section 6 above). This is the so-called circular velocity of the satellite. If f and m are equal to: $f = 6.67 \times 10^{-8}$, $m = 5.974 \times 10^{27}$ gr. then $\sqrt{fm} = 1.99 \times 10^{10}$. A more precise determination of this quantity, based on a study of the earth's shape and the acceleration due to gravity at the earth's surface, gives in c.g.s. units $\sqrt{fm}=1.99654\times10^{10}$.*

Taking this value, we obtain for the circular velocity the formula

$$V_0 \text{ cm/sec.} = 1.99654/r_0$$

where r_0 is expressed in centimetres. To facilitate computations it is best to transform this expression. To do this, we first calculate the circular velocity V_{00} of a hypothetical satellite moving in a circle with a radius equal to the equatorial radius of the earth $R=6,378$ km., that is, at the very surface of the earth.**

If in the latter equation we put $r_0=R=6,378\times10^5$ cm, we will obtain $V_{00}=7,906\times10^5$ cm/sec. $=7,906$ m/sec. After multiplying and dividing by $\sqrt{R}$ we rewrite the formula for V_0 as follows:

$$V_0 = \frac{1.99654\times10^{10}}{\sqrt{R}}\sqrt{\frac{R}{r_0}} = 7,906\sqrt{\frac{R}{r_0}}\ \text{m/sec.}$$

This formula is convenient in that it contains only the ratio between the equatorial radius of the earth R and the distance r_0 of the satellite from the earth's centre.

The circular velocity of the satellite diminishes with increasing r_0. For example, at a height of 100 km. above the earth's surface near the equator (i. e., at $r_0=6,478$ km.) we have $V_0=7,845$ m/sec., at 300 km. ($r_0=6,678$ km.) $V_0=7,727$ m/sec.

If the initial velocity V_0 of the satellite is greater than the circular velocity but less than the parabolic velocity the orbit will be elliptic. This has been the case in the launching of all artificial satellites so far. The eccentricity of this orbit is connected with the velocity v and v_0 by the relation

$$e = \left(\frac{v}{v_0}\right)^2 - 1$$

---

* Obtained from material in K. A. Kulikov's book *Fundamental Astronomical Constants*, Gostekhizdat, 1956.
** To make this more explicit, note that if the satellite's orbit passed along the equator it would lie at the very surface of the earth. But if it passed perpendicular to the equator it would rise 21 km. above the earth's surface at the north and south poles.

while the semi-major axis of the orbit (or the mean distance of the satellite from the centre of the earth) is

$$a = \frac{r_0}{1 - e}$$

In this case, the perigee (closest point to the earth's centre and, hence, to the surface of the earth) will be directly above the launching site of the satellite. The apogee (most distant orbital point from the centre and surface of the earth) will be on the exact opposite side of the globe. These points of course refer to the respective positions of the satellite above the earth's surface at the initial instant of time. In the course of time, the perigee and apogee of an unperturbed orbit will retain the same position in space but will be in motion with respect to geographical points on the earth. This is because the earth rotates on its axis, while the unperturbed satellite orbit does not alter its position in space. When the satellite circles the earth once, the latter has turned through a certain angle so that the satellite's next circuit passes over different parts of the earth.

The apogean distance of the satellite from the centre of the earth is

$$r_A = a\,(1 + e)$$

and the apogean height above the earth's surface is

$$h_A - r_A - 6{,}378$$

The perigean distance r_p from the earth's centre and the perigean height h_p above the earth's surface are related to a and e by the equations

$$r_p = a(1 - e) \text{ and } h_p = r_p - 6{,}378$$

Here, a is throughout expressed in kilometres.

It should be noted that if the height above the earth's surface is calculated as the difference $(r - 6{,}378 \text{ km})$., where r is the distance to the centre of the earth, the true height will be only that above the equator. For the other points on the earth there will be a discrepancy due to the earth's compression, that is, due to the fact that the distance from the centre to the surface is less than 6,378; at the poles this difference amounts to 21 km.

The orbital period T_0 of a hypothetical satellite moving in a circular orbit at a distance of 6,378 km. from the earth's centre is:

$$T_0 = \frac{2\,\pi R}{V_{00}} = \frac{2\pi \cdot 6.378 \cdot 10^3}{7,906} = 5,069 \text{ sec.} = 84.48 \text{ min.}$$

According to Kepler's Third Law (see Section 2) the orbital period of a satellite moving in an elliptical orbit with a semi-major axis a should satisfy the relation:

$$\frac{T^2}{T^2_0} = \frac{a^3}{R^3}$$

or

$$\frac{T^2}{(84.48)^2} = \frac{a^3}{6,378^3}$$

where a is expressed im kilometres and T in minutes of time.

Through observations of a satellite during the first days of its existence we can compute its orbital elements and determine to what extent the orbit differs from the prescribed one. All three of the Soviet satellites were successfully launched into their prescribed orbits.

Immediately after launching, the first Soviet Sputnik had an orbital period of 96.15 min. (96 min. 9 sec.). From the equation relating T and a we find that this period corresponds to the following semi-major axis of the orbit;

$$a = 6,378 \left(\frac{96.15}{84.48} \right)^{2/3} = 6,953 \text{ km.}$$

If, in addition, we know h_A and h_p, we can calculate the eccentricity of the orbit from the other equations:

$$e = \frac{h_A - h_p}{2a}$$

and in this way we will determine the complete size and shape of the orbit. For Sputnik I, the initial h_A and h_p were, according to reports in the press, 947 and 228 km. respectively.

From this it follows that

$$e = \frac{719}{13,906} = 0.0517*$$

The orbital plane of Sputnik I was inclined to the plane of the equator at an angle of $i = 64°.26$. Fig. 40 gives the general position of the orbit relative to the earth.

Following are the initial orbital elements of Sputnik II and Sputnik III:

II	III
$T = 103.75$ min.	$T = 105.95$ min.
$a = 7,314$ km.	$a = 7,418$ km.
$e = 0.09885$	$e = 0.113$
$h_p = 225$ km.	$h_p = 224$ km.
$h_a = 1,671$ km.	$h_a = 1,880$ km.
$i = 62°.5$	

These may be compared with the following orbital elements of the first and second American earth satellites:

---

* It is also possible to calculate the eccentricity from the period T and either the height h_A or h_p. True, in this case one must know rather accurately the distance R from the surface to the centre of the earth at the point of observation. For example, for Sputnik I we take $h_A = 947$ km., $R = 6,378$ km. Then, using the above equations we obtain

$r_A = 6,378 + 947 = 7,325$ km, $e \frac{r_A}{a} - 1 = 0.0535$, $r_p = 6,581$ km. $h_p = 203$ km.

e and h_p will be found to differ from the earlier computations. This is because we incorrectly took the distance to be $R = 6,378$ km.

This figure is correct for points on the equator, but the satellite's apogee, apparently, was not above the equator. The distance to the centre of the earth diminishes as we move away from the equator. Since in this individual case we know both h_A and h_p it is possible to calculate that at perigee and apogee the distances R_A and R_p to the earth's centre from the surface were, during the first few days, approximately 6,365 and 6,366 km. Indeed, on the basis of the above values $a = 6,953$ km. and $e = 0.0517$ and $h_A = 947$ km., $h_p = 228$ km., we obtain

$r_A = a (1+e) = 7.312$ km., $r_p = a (1-e) = 6,594$ km.,

and therefore

$R_A = r_A - h_A = 6,365$ km., $R_p = 6,366$ km.

If in calculating h_p from the given period T and h_A we assumed $R = 6,357$ km. (the distance from the surface to the centre at the poles), we would obtain $h_p = 245$ km. The values of h_p calculated from $R = 6,378$ km. and $R = 6,357$ km. would then differ by 42 km. To summarize, h_p may be computed from h_A and T (h_A from T, h_p) if very approximate values are needed or when the distance R is known with sufficient accuracy.

	I		II
T	$=114.95$ min.	T	$=135$ min.
a	$=7,831$ km.	a	$=8,731$ km.
e	$=0.14052$	e	$=0.2$
h_p	$=352$ km.	h_p	$=650$ km.
h_a	$=2,554$ km.	h_a	$=4,000$ km.
i	$=33°58$	i	$=33°$

The terrestrial atmosphere plays a very essential role in the motion of artificial earth satellites. Due to the resistance of the air, the satellite is decelerated and gradually comes closer to the earth. At heights of 100-150 km. above the earth's surface, air resistance is such* that the satellites heat up, break into pieces and fall to earth like ordinary meteorites.

The first Soviet satellite stayed in orbit till the beginning of January 1958, or about 3 months, completing a total of about 1,350 circuits of the earth. On December 31 it was reported to have a period of about 90 min. and an apogee of 320 km. Using published data we compiled a table indicating the orbital periods and semi-major axis of Sputnik I for different dates. Also included in the table are the values of e, h_p, h_a for instants at which it was possible to calculate them.

Date	T, min.	a, km.	e	(h_a) km.	(h_p) km.
4 Oct. 1957	96.15	6,953	0.0517	947	228
9	96.02	6,946	—	—	—
21	95.55	6,924	—	—	—
25	95.44	6,918	0.0460	890	220
27	95.31	6,912	—	—	—
9 Nov.	94.72	6,884	0.0442	810	202
31 Dec.**	~ 90			~ 320	

From this table we may draw the following conclusions.
1) The satellite gradually came closer to the earth, the

---

* Even though the air density at these heights is less than that at the surface of the earth by a factor of millions.
** There is a mistake here in either the period or the apogee. For these values of T and h_a we obtain $a=6,660$ km. and $h_p=244$ km. which cannot be since h_p must diminish with time. If we retain $h_a=320$ km., it is more likely that $T=89$, in which case h_p would be 150 km.

semi-major axis of its orbit a, and the heights h_a and h_p decreased in time. And the orbital period of the satellite likewise diminished —a very paradoxical thing at first glance: the atmosphere impedes the motion of the satellite, yet the latter circuits the earth faster and faster!

It may be noted that even the mean velocity of orbital motion increases. Calculating the length of an arc of the ellipse by equation $S = 2\pi\,a(1-1/4\,e^2-\frac{3}{64}e^4)$, which holds for small values of eccentricity, e, we find that on Oct. 4, 25 and Nov. 9 the paths covered by the satellite during each circuit about the earth came out to 43,696 km., 43,444 km., and 43,232 km. respectively. Dividing these by the orbital period (in seconds) we obtain the mean velocities: 7.574 km/sec., 7.587 km/sec. and 7.607 km/sec., respectively.

2) T and a diminish at an accelerated rate. Between Oct. 4 and 21 the average rate of change of T was 2.12 seconds per 24 hours, between Oct. 21 and Nov. 9—2.62 seconds per 24 hours and between Nov. 9 and Dec. 31—about 5 seconds every 24 hours. The mean daily variation of a during the periods Oct. 4-21 and Oct. 21-Nov. 9 was 1.82 km. and 2.11 km., respectively.

3) The eccentricity of the orbit diminished with the semi-major axis, which means that the orbit became less and less elongated.

4) The perigee distance, h_p, diminished much more slowly than did the apogee distance, h_a. Between Oct. 4 and Nov. 9, h_p decreased by 26 km., while h_a lost 137 km.

Circling about the earth along with Sputnik I (the satellite proper) was the carrier rocket*—itself an independent satellite. At the beginning the carrier rocket was not far from the satellite. But since it encountered greater atmospheric resistance than did the satellite proper (this was mostly due to its shape), it began to lose height faster than its companion. And in doing so it did not lag behind Sputnik I but overtook it since its orbital period diminished more rapidly. It lived up to Dec. 3, 1957, and during the last three days of its existence it moved in the relatively dense layers of the atmosphere and glowed giving the impres-

---

* Or, to be precise, the last stage of the composite rocket that launched Sputnik I into orbit.

sion of burning. On Dec. 1 it was observed over Riga at 7 a. m. as a bright yellow ball of fire. During the last days of November its orbital period was about 90 min.

The orbital elements of the carrier rocket (T, a, e, h_a, h_p) varied in the same way as those of the satellite proper. Below is a table, similar to the one above, for the rocket carrier computed from published data:

Date	T, min.	a, km.	e	(h_a), km.	(h_p), km.
4 Oct. 1957*	96.15	6,953	0.0517	947	228
21 Oct. 1957	95.12	6,903	—	—	—
25 Oct.	94.68	6,881	0.0416	789	217
9 Nov.	93.48	6,823	0.0366	695	195

The second Soviet satellite had a longer lifetime than the first; it lasted till April 14, 1958, or about four and a half months, completing a total of 2,370 circuits about the earth.

The following table for Sputnik II contains data on T, a, e, h_a, and h_p.

Date	T, min.	a, km.	e	(h_a), km.	(h_p), km.
3 Nov. 1957	103.75	7,314	0.0988	1,671	225
9 Nov.	103.52	7,303	—	—	—
17 Dec.	101.59	7,214	0.0876	1,468	204
31 Dec.	100.76	7,173	—	—	—
28 Jan. 1958	98.87	7,083			

All the foregoing variations in the orbital elements of artificial satellites may be explained theoretically.

Since satellites move in elongated orbits and the atmospheric density falls off rapidly with distance from the earth's surface, the resistance of the air is greatest near the perigee of the orbit, in other words, when closest to the earth. At apogee, the satellites experience hardly any deceleration at all. Schematically, satellite motion may be represented as follows. Let the initial orbit be ellipse 1 in Fig. 41. The dashed line is the boundary, Q, above which atmospheric resistance is too small to be taken into account.

Resistance due to the atmosphere is perceptible only

---

* From the very beginning the velocity of the carrier rocket was less than that of the satellite proper, since the latter was ejected from the last stage of the rocket (true, the thrust was only slight). This is why the figures in this line are not exact. All the quantities—a, T, h_a and h_p should be reduced somewhat.

over the portion of orbit close to perigee, P. Let us denote by V_1 the velocity of the satellite emerging from perigee, P, needed in order to move in ellipse 1. As the satellite recedes from perigee its speed diminishes and reaches a minimum at apogee. Here it is equal to

$$V_1 = \frac{\sqrt{fm} \times \sqrt{1 - e_1}}{\sqrt{a_1 (1 + e_1)}}$$

where a_1 is the semi-major axis, and e_1 is the eccentricity of ellipse 1.

When the satellite returns to perigee P in orbit 1 it experiences retardation. If we regard the satellite as being decelerated only at P (see Fig. 41) its direction will be retained as it emerges from perigee but it will have less velocity. We designate the new velocity by V_2 $(V_2 < V_1)$. The next circuit will now lie in orbit 2, for which the apogee distance h_a and the semi-major axis a_2 will have diminished.* The greater the density of the air near P the more speed the satellite will lose and the faster h_a and a will diminish.

The velocity V_2 will not be sufficient now for the satellite to overcome the earth's gravitation and recede to point A_1. Orbit 2 is less elongated and its eccentricity e_2 is less than that, e_1, of ellipse 1. At the apogee of orbit 2, the satellite will have a velocity,

$$V^1_2 = \frac{\sqrt{fm} \times \sqrt{1 - e_2}}{\sqrt{a_2 (1 + e_2)}}$$

And since $a_2 < a_1$ and $e_2 < e_1$, the velocity, V^1_2, is greater than V^1_1. By Kepler's Third Law we again find that the orbital period in orbit 2 is less than that of orbit 1.

Thus, although the satellite was retarded near perigee

---

* All proportions in Fig. 41 are greatly exaggerated so as to facilitate explanations. First, the orbits are too elongated and the apogee is too far away from the earth (upwards of several earth radii). Actually the apogees of the first Soviet satellites were at distances less than an earth radius. Second, a single circuit of the satellite does not produce such a reduction in the apogee distance. For example, between Oct. 4 and 25 Sputnik I lost an average of 2.7 km. in apogee distance per 24 hours, while in a single circuit (the satellite did 15 circuits a day) this distance amounted to only 0.18 km., which is roughly 1/5,000 of the original apogee distance.

and lost speed, near apogee it gained speed. This is because at apogee the satellite was now closer to the earth and was experiencing a greater pull by the latter. Now an increasing gravitational force leads to higher accelerations and velocities of the satellite. The retardation effect near perigee is balanced by the increasing gravitational pull of the earth

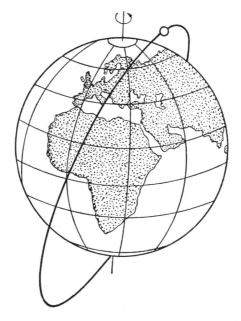

Fig. 40. The orbit of the first artificial earth satellite.

near apogee, and the resultant mean velocity of motion of the satellite in its orbit increases.

Above, we started from the rough approximation that retardation occurs only at perigee P, in which case, the apogee distance will gradually diminish while the perigee distance will remain constant. In reality retardation occurs over a certain portion of the orbit close to perigee (Fig. 41). This brings about a reduction, though relatively slight, in the perigee distance. The satellite will return not to point P but to P_1, which is closer to the earth. The

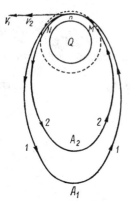

Fig. 41. Retardation of air density at great heights above the satellite in the atmosphere earth. The satellite loses height the (schematic).

result is that the satellite descends into denser layers of air and retardation at perigee gradually increases. And this leads to a gradual decrease in the semi-major axis of the orbit and the period of revolution of the satellite.

Such is the disturbing effect of the terrestrial atmosphere. We see that the perturbations are secular (see Section 13), that is, the orbital elements a, T, e vary in one direction. The study of these perturbations is of great significance in determining the air density at great heights above the earth. The satellite loses height the faster, the greater is the retardation and, along with it, the density of the air near the perigee of the orbit (at the beginning). The rate of descent of a satellite will indicate the amount of retardation, and hence also the density of the atmosphere. The first results obtained from observations of the first two Soviet satellites have shown that the air density at 200 km. is from 5 to 10 times greater than had been supposed.

Perturbations of satellites due to the earth's compression and to the uneven density of material in the earth's interior do not lead to a constant decrease or increase of the semi-major axis, the eccentricity and the orbital period. These perturbations do not, on the average, alter the size and shape of the orbit, but gradually turn it in space. Studies of these perturbations in the case of real satellites can help to define more accurately the shape of the earth and also to extend our still insufficient knowledge concerning the density distribution of material in the interior.

17. THE MOTIONS OF ASTEROIDS

We have already mentioned that close to 1,600 asteroids have been registered to date. By registered is meant that the asteroid's orbit has been derived at least approximately. The number of discovered but not registered asteroids, however, is far greater than 1,600, with new ones constantly being discovered and their orbits computed.

Since there are already a large number of asteroids and more are being discovered all the time, it is necessary to keep tab on them by means of something in the nature of an "asteroid patrol" in order to be able to study their motions and to distinguish them from newly discovered bodies. It is naturally practically impossible to keep all the asteroids under daily observation so as not to lose sight of them. Ordinarily, known asteroids are observed only when they make a close approach to the earth (at opposition with the sun). Many of the asteroids, due to smallness of size, can be seen only at this time. For this reason, it is necessary to know exactly how each asteroid moves in space and to be able to compute its future positions in the sky for any instant of time. From time to time, observations are made and the apparent positions in the sky are determined, thus permitting a check to be made on how correctly we calculate these positions. But this requires a theoretical calculation of the motion of the body in space, which is done by applying Newton's law of gravitation.

It is evident, therefore, that theoretical investigations of the motions of asteroids are a prerequisite to asteroidal observations proper. Besides, there has recently appeared a practical necessity for very precise theories of motion of certain of the minor planets. These observations should help to refine the so-called *fundamental astronomical constants** which are necessary in the compiling of star catalogues.

The problem of asteroidal motion is at once simpler and more complicated than that of planetary or satellite motion. Asteroids have very small masses and so exert such a weak influence on the motions of other bodies that it has as yet escaped our notice. For this reason, when investigating the motions of asteroids one can disregard both their mutual attractions and their gravitational pull on the major planets. Asteroids are regarded as moving under the influence solely of the gravitation of the sun—which acts as the centre of force—and of the disturbing attraction of the planets. Further, the planets are regarded as moving in definite or-

---

* Fundamental constants are quantities that describe the shape and size of the earth, and its motion and distance from the sun.

bits that do not in the least depend on the motions of the asteroids. This, of course, greatly simplifies the problem.

The first task following the discovery of a new asteroid is to derive its unperturbed elliptical orbit. This can be done if at least three apparent positions of the asteroid have been observed at instants of time separated by intervals of several days. However, it often happens that a newly discovered asteroid is lost to sight before these data have been obtained. Sometimes observations are impossible due to overcast skies. In this way, many discovered asteroids are "lost" and remain unregistered. Between 1911 and 1930 a total of 1,962 asteroids were discovered, but only for 484 were the orbits computed and thus only these were registered. Naturally, the first orbit derived is not especially accurate, and further observations are required to make it more exact.

But one cannot restrict himself to determining the elliptical orbit of an asteroid, for its perturbations are often very great. In practice, only the disturbing forces of Jupiter and Saturn are taken into account, since the gravitational forces of these planets are most noticeable. In constructing theories of the motions of asteroids, wide use is made of the method of successive approximations.

However, asteroidal motions are far more complex than planetary motions. As a rule, the elliptical orbits of the asteroids are much more elongated than planetary orbits, and the planes of motion of the majority of asteroids are inclined at greater angles to the plane of the earth's motion than are the planetary orbits. This, above all, gives rise to purely mathematical difficulties in calculating the perturbations of the asteroids. Further, many asteroids make close approaches to Jupiter and can experience rather strong Jovian attraction. Fig. 42 shows the orbits of Jupiter and the asteroids Pallas and Juno*. The orbits of Pallas and Juno have semi-major axes of 2.8 A. U. and 2.7 A. U. respectively, and eccentricities of 0.24 and 0.26. Juno can approach Jupiter to nearly 1.9 A. U., while Pallas can come to a distance of roughly 2 A. U. if we take into account the angle between the planes of the orbits of Jupiter and Pallas. In these positions, the solar attraction exerted on these as-

---

* The figure does not show the orbital inclinations of the asteroids.

teroids will exceed the gravitational pull of Jupiter by only 290-300 times.

For this reason, the perturbations of the asteroids are far greater than those of the major planets, and even over relatively short periods of time are reckoned not in seconds or minutes of arc but by tens of minutes and even degrees. Again, this is a factor complicating the accurate computation of perturbations.

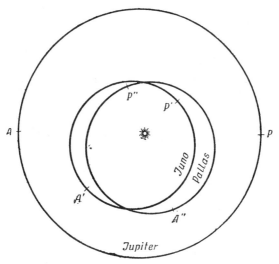

Fig. 42. The orbits of Juno and Pallas (*A* and *P* indicate the aphelia and perihelia of the orbits).

For this reason and maybe also because astronomers have paid far less attention to asteroids than to the major planets, theories of motion have been worked out for only a few asteroids, and the accuracy of these theories is considerably below that of the theories of motion of the major planets. At any rate, present-day observations of asteroidal positions in the sky are made with an accuracy up to $1''$, while divergences between observations and the best theories of asteroidal motion reach tens of seconds of arc.

At the present time, numerical methods are widely used in computing perturbations and compiling tables that indicate the apparent positions of asteroids in the sky and other data of asteroidal motion. These methods enable one to indicate the positions of asteroids in space and their apparent positions in the sky with a far greater degree of accuracy than do analytical techniques. This is the method used in the Institute of Theoretical Astronomy, U. S. S. R. Academy of Sciences, in Leningrad to compile *The Ephemerides of the Minor Planets*. This volume contains data on the motions of all registered asteroids. The information collected here is sufficient for tracking the motions of the asteroids.

18. PLANETARY ROTATION

Up till now we have spoken of the motions of the planets and other bodies around the sun and of satellites around their primaries. This is the motion of a body in revolution, it is determined by the motion of its centre of gravity. But in addition to this *translational* motion, celestial bodies are also in motion relative to their centre of gravity—*rotational* motion.

The rotation of the earth about a certain imaginary line, which we call the axis of rotation, is illustrated by the diurnal rotation of the entire celestial sphere. The earth's rotation is also detected by means of certain physical experiments (Foucault's pendulum experiment), and is confirmed by a large number of phenomena observed on the earth*. The earth's period of rotation as determined from stellar observations is equal to 23 hours 56 minutes and 48 seconds.

The rotation of Mars, Jupiter and Saturn can easily be detected by careful observation in a telescope; various spots and other surface markings that are visible telescopically do not remain in one place but appear to move over the disk from one limb to the other and then disappear from view. This is evidence that the planet is in rotation.

Observations of this kind help to determine the rotational periods of these planets. Mars has a period of approximately 24 hours 37 minutes, Jupiter's period of rotation is

---

* For instance, the earth's rotation explains the fact that rivers in the Northern Hemisphere have steep right banks and sloping left banks, whereas rivers in the Southern Hemisphere have sloping right banks and steep left banks.

9 hours and 50 minutes, and Saturn's period is 10 hours and 20 minutes. No surface features have been detected on the other planets, whose rotational periods have been determined by means of special observational methods. Uranus has a period of 10 hours 45 minutes, while Neptune's period is about 15 hours. Mercury moves about the sun so that one side of the planet is always facing the sun, which means that its period of axial rotation is equal to its orbital period, that is, roughly 88 days. Our data concerning the rotational periods of Venus and Pluto are as yet unreliable.

Observations show that the planets rotate in such a way that their axes of rotation retain a constant *orientation* in space for very long periods of time. Their orbital periods also remain constant.

We can observe directly the rotational motion of the moon—the earth's satellite. The moon always keeps one side turned towards the earth, which means that it rotates once on its axis during the same time that it completes a single circuit about the earth (about 27 days).

Although special methods permit detecting rotational motion of many bodies of the solar system, of greatest interest is the rotational motion of the planets and the moon. We shall deal here only with planetary rotation.

There is of course no sense in asking why the planets rotate or why they have motion in general. Everything in the universe, from the smallest dust particle to colossal cosmic bodies, is in constant motion. There is no such thing as matter without motion. The matter that later went to form the planets was also in motion. In the process of their "birth" the planets acquired both translational orbital motion about the sun and rotational motion* on their axes.

However, we can and should ask how the planets move, what forces govern these movements, and what laws of motion the planets obey. We have already analysed their orbital motions. Let us now examine the question of the rotational motion of the planets.

According to the basic laws of mechanics, if a body is not acted on by any forces it should move by inertia. If the body had translational motion it should, in the absence of forces, be at rest or should move uniformly and rectilin-

* For details see *The Origin of the Earth and the Planets* by B. Levin, Foreign Languages Publishing House.

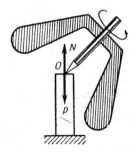

Fig.43. The Maxwell top.

early with the same velocity that it had at the initial instant. In the general case, rotational motion by inertia (that is, rotation in the absence of acting forces) appears far more complex. Let us examine a special case of this motion, which is related directly to the problem of planetary rotation.

If a *solid* body with an axis of symmetry (for example, a cone, cylinder or ellipsoid of revolution) is in rotation about this axis, the rate of rotation of the body and the orientation of the axis of rotation in space should, in the absence of forces, remain constant. This is what we find in the case of the Maxwell top pictured schematically in Fig. 43.

The Maxwell top is bell-shaped with heavy sides. Its centre of gravity is inside the bell at point O. The axis, on which the bell is mounted, ends in a sharp point at the centre of gravity O. If the top is placed so that the point O rests on the support, the force of gravity of the top P will manifest itself in the pressure of the top on the support and will be balanced by the resistance of this support N.

In other words, the top will be a body on which no forces (which do not balance each other) are acting, and for this reason it should move by inertia. If given a certain position, it should retain this position and remain at rest. If we push the axis of the top slightly, it will oscillate a bit and then settle into a new position.

Now let us spin the top on its axis. It will spin and retain the position that it was given on the support. And what is more, if the top is in fast rotation, quite a considerable additional force is required to alter the direction of its axis.

The same things will occur if the top is in no way supported but is simply let to fall in the air, in which case the top will be acted upon by the force of gravity that will make it fall. The direction of this force passes through the centre of gravity of the top. If before being thrown up the top is given a strong twist on its axis it will continue to spin with the same velocity and its axis will all the while retain one and the same direction (Fig. 44).

Now let us lengthen the axis of the top, put it on the support and give it a strong spin. The top is now acted upon by the force of gravity and the resistance of the support which do not balance each other and strive to turn the top over. If the top were not spinning it would simply fall down. But it does not fall when rapidly rotating. Instead, the axis of the top oscillates regularly describing a cone about the vertical line (Fig. 45). The rate of this spin remains constant.

We encounter a similar situation when considering the rotational motion of the planets. We shall give one example,

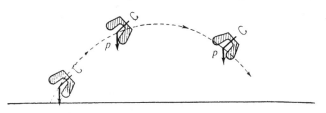

Fig. 44. The axis of rotation of a rotating top thrown upwards remains constant.

that of the rotational motion of the earth, since the latter has been studied in greatest detail and, practically speaking, is the most important. The approach to the study of planetary rotational motion is essentially the same.

Like the other planets, the earth's shape resembles that of an ellipsoid of revolution with a slight compression along the axis of rotation (the equatorial radius is greater than the polar radius). For this reason, the earth cannot be attracted by other heavenly bodies exactly like a sphere, and the forces of attraction acting on the earth do not pass exactly through the latter's centre of gravity.

We have already considered the law for attraction by a spheroid. From this law it follows that the force with which the earth is attracted by some other heavenly body, let us say M (Fig. 46), differs from the force with which M would attract a sphere above all in magnitude. Besides, it does not only impart to the earth a translational motion but also strives to turn the earth's axis of rotation. This is clearly seen from Fig. 46. Calculations show that the magnitude of the force that strives to turn the earth's axis of

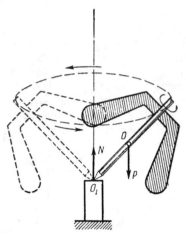

Fig. 45. The precession of a top.

rotation is proportional to the mass, m, of the attracting body M and is inversely proportional to the cube of the distance of this body from the centre of the earth. Comparing the forces acting on the earth and the forces acting on the top, we can conclude that the forces of attraction of different heavenly bodies with respect to the earth should not affect the rate of the earth's rotation but should lead to regular oscillations of the earth's axis of rotation. What bodies should exert the greatest influence on the earth's rotation?

First, the moon since it is closest to the earth; second, the sun, which though farther away has by far the greatest mass. The planets exert a very weak influence because the perturbing action rapidly diminishes with distance. From the relationship $F \sim m/r^3$ we find that the influence of the sun is exceeded by that of the moon 2.2-fold; and the moon's action is 13,000 times that of Venus, 140,000 times that of Jupiter, and 800,000 times that of Mars. The effect of the remaining planets on the earth's rotation is still weaker.

The earth moves about the sun, while the moon circles the earth. For this reason, the moon-earth-sun configuration is continuously changing, with the result that the magnitude and direction of the forces acting on the earth's rotation axis are also continuously changing. Due to these forces, the earth's axis of rotation describes a complex motion. First of all it slowly describes a cone remaining all the time inclined to the plane of the earth's motion at an angle of 23°.5 (Fig. 47). This is the so-called *precessional motion* of the axis of rotation, which determines its mean direction in space at different epochs. This motion has a period of roughly 26,000 years. In addition, the earth's axis of rotation describes various slight oscillations relative to its

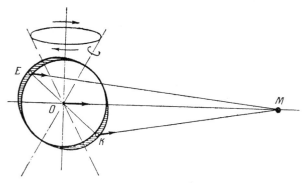

Fig. 46. An explanation of terrestrial precession resulting from lunar attraction.

mean position. Most important of these oscillations is the so-called principal *nutation* with a period close to 19 years.

How can one detect such a motion of the earth's axis of rotation? Why is a study of this motion so important?

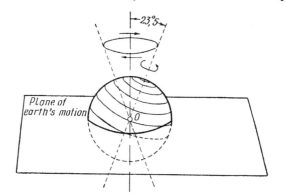

Fig. 47. The precessional motion of the earth's axis.

We observe celestial bodies from the earth's surface. Therefore, their positions can only be determined with respect to reference points associated with the earth. Such reference points are the plane of the earth's motion about the sun (the plane of the ecliptic), the direction of the earth's axis of rotation and the plane of the equator of the earth perpendicular to this axis. The direction of the rotation axis intersects the celestial sphere at the north and south poles of the world, about which poles the celestial sphere produces its apparent diurnal rotation. These reference points, which are connected with the earth but do not participate in the earth's diurnal rotation, permit one to "fix" each heavenly body in place on the celestial sphere.

Due to precession of the earth's rotation axis the poles describe in the sky circles of radius about 23°.5. Since the poles make a complete circuit in just about 26,000 years, in one year they move approximately 50'. The nutational oscillations of the earth's rotation axis lead to additional periodic displacements of the poles that reach 9".

Alterations in the direction of the earth's axis of rotation lead to a change in the plane of the equator and to an altered position of the poles of the world in the sky, that is to say, to an apparent shift of the celestial bodies in the sky relative to these reference points. These apparent shifts due to the earth's motion are compounded with those caused by the actual motion of the heavenly bodies in space. We will get a correct picture of the actual motion of the heavenly bodies only when we find out what part of their apparent motions in the sky is due to their actual motion in space and what part is due to the earth's motion. This is why, when studying the motions of the heavenly bodies, we must know how these reference points change.

The stars are at great distances from the earth and retain almost unchanged their position in the celestial sphere. Changes in the stellar positions relative to the poles of the world and the plane of the equator and the ecliptic reflect primarily not the motions of the stars proper but the displacement of the poles, the equator and the ecliptic. It was precisely this alteration in the stellar positions that was detected as far back as 2 000 years ago by the Greek astronomer Hipparchus, who called this phenomenon by the name *precession*. But at that time the cause of it was not understood. It was only at the beginning of the eighteenth century that Newton, applying the law of gravitation, explained precession and predicted nutation and other oscillations of the terrestrial axis. Nutational oscillations of the positions of stars (stellar *nutation*) was discovered only in the middle of the eighteenth century.

The rotational motion of the earth was studied on the basis of stellar observations and also theoretically. However, at the present time we need more accurate determinations of the precessional and nutational motions of the earth's rotation axis. The point is that the stars are moving in space like all other bodies, and the nature of these motions is still rather vague. Which means that there is not much sense in determining the movements of the poles and the plane of the equator from the observations of stars, that is, objects whose motions have yet to be defined.

In order to determine with more accuracy the magnitudes of precession and nutation we must observe objects whose positions in space and the sky are well known, even though they may not be permanent. Such objects are the planets and asteroids of our solar system; their motions in space about the sun are studied by means of Newton's law of gravitation. We already have accurate theories of the motions of the major planets. However, in practice it is more convenient to use the asteroids because they appear as luminous points, whereas the planets exhibit noticeable disks. The positions of stars and asteroids (these are necessary in determining the constants of precession and nutation) can be correlated with far greater accuracy than the positions of stars and the major planets. This is why we need precise theories of motion of the asteroids, as was mentioned in Se tion 15.

We exam:ned the motion in space of the earth's axis of rotation caused by the perturbing action of the moon and sun. But how does the earth rotate on this axis?

If we take into account only the forces of mutual attraction between the earth and the other heavenly bodies, the rate of the earth's rotation should not change at all. Our day, determined by the period of rotation of the earth, should remain constant. Yet, in reality the rate of the earth's rotation is gradually diminishing. The terrestrial day is very slowly increasing by 0.001 second per century. How is one to account for this increase in the day?

The change in the length of the day is connected with the tides. Everyone knows that the level of water in the oceans does not remain the same throughout the day but changes regularly. In the course of six hours it rises reaching a maximum at *high tide*; during the next six hours the water ιecedes and reaches a minimum at *low tide*, and so on in ιegular alternation.

In localities located on the same meridian the tides occur at almost one and the same time; to the east of the given locality they set in earlier, and to the west, later. Thus, a *tidal wave* extending the length of the meridian is in motion in the oceans from east to west, that is, in a direction coun-

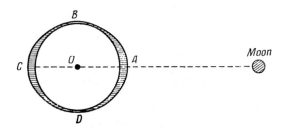

Fig. 48. The formation of a tidal wave.

ter to that of the earth's rotation. At the equator, this wave moves at a speed of 1,600 km/hr. and encircles the globe in 24 hours 50 minutes. Such also is the duration of the apparent revolution of the moon about the earth. This fact long ago suggested the idea of some sort of connection between the tides and the moon. But Newton was the first to give a correct explanation of the tides on the basis of the law of universal gravitation.

The attraction of the earth by the moon consists in the attraction by the moon of the individual particles that comprise the earth. The particles which at the given moment are closer to the moon experience greater attraction, those that are farther away are more feebly attracted. If the earth were rigid, this difference in attraction of the individual particles would not matter. Then one could speak of the movement of the earth as a whole, as determined by the motion of the earth's centre and the rotation of the earth on its axis. However, the earth is not absolutely rigid and, besides, it is covered with oceans (71% of the terrestrial surface is occupied by water), and particles of liquid will move in a slightly different way from that of the rigid part of the earth.

The particles of water closest to the moon at the given instant (near point A in Fig. 48) and farthest away (near point C) are attracted by the moon with forces that differ. Particles at A are attracted by the moon more strongly, and particles at C weaker than particles at the centre of the earth. The result is that the particles of water near A are pulled towards the moon more, and particles near C less

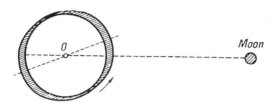

Fig. 49. Displacement of the tidal wave due to the earth's rotation.

than the centre O and the entire rigid body of the earth. On the moonward side of the earth the water will rise producing a high tide. On the opposite side, at C, there will also be a high tide, since the particles of water near this point will lag behind the earth's centre in its motion towards the moon. Thus, near A and C there will be high tides with an excess accumulation of water, and at B and D the level of water will fall, producing a low tide.

The tidal bulges near A and C strive to maintain one and the same position with respect to the moon. But with respect to the earth the tidal bulges alter their positions due to the earth's rotation and move in a direction counter to the earth's rotation. In its motion from east to west the tidal wave experiences friction over the ocean floor and resistance offered by the continents that it encounters. For this reason the rotating earth pulls the tidal bulges along and they occupy, with respect to the moon, positions shown in Fig. 49. This is why the tides in each locality lag behind the instant of meridian passage of the moon (either at upper or lower culmination).

The force of friction between the tidal wave and the ocean floor, and also the forces of internal friction due to viscosity of the water retard the earth's rotation. But tidal waves are formed not only in the water envelope of the earth but in the rigid body too because the earth is not perfectly rigid. These tidal waves also move through the earth due to the rotation of the latter, and the internal friction thus produced, due to the viscosity of the terrestrial matter, likewise slows down the earth's rotation. The overall result is that the rotation of the earth is gradually decelerated. We have already said that the terrestrial day increases roughly 0.001 second per century.

How is it possible to detect such a minute alteration in the length of the day? This is of course very difficult in direct observations of the length of the day, but for considerable periods of time the influence of an increasing day can be detected.

Let us suppose that one clock follows the earth's rotation exactly, while another clock does not lose time at all. For the sake of simplicity, let us assume that at the present time the earth's period of rotation is equal to one mean solar day (in place of 23 hours 56 minutes of mean solar time). In 100 years our day, that is, the time of one full circuit of the hour-hand of the first clock,* will decrease by 0.001 sec., and each day during these 100 years this time will diminish by $\varepsilon = 0.001/36,525$ seconds (100 years $= 36,525$ days). Thus, if the daily angular rate of rotation of the hand (or, in other words, the angle swept out by the hand in hours, minutes or seconds of time** during one day) always remains constant for the second clock and equal to ω_0, it will be constantly varying for the first clock and will be equal to $\omega = \omega_0 - \varepsilon t$, where t is expressed in days. The quantity ε is the retardation of the angular rate of rotation of the hand of the first clock or the slowing down of this clock. The number of circuits during time t for the second clock is equal to $T_0 = \omega_0 t$ and for the first clock is (according to the equation for uniformly retarded motion) equal to $T = \omega_0 t - \dfrac{\varepsilon t^2}{2}$ The time lag $T_0 - T$ between the first and second clock during time t will amount to $T_0 - T = 1/2 \, \varepsilon t^2$, where t is ex-

* The clock dial is divided into 24 hours, not 12, so that the hour-hand completes one circuit a day.
** In astronomy, angles are often expressed in units of time: 360° corresponds to 24 hours or 1,440 minutes or 86,400 seconds of time.

pressed in days. In 100 years, that is, approximately 36,525 days, the time lag of the first clock will amount to about

$$1/2 \times 0.001 \times 36,525 \approx 18 \text{ seconds};$$

or 1,800 seconds (0.5 hr.) in 1,000 years and nearly 2 hours in 2,000 years.

Astronomy makes use of time reckoned from the earth's rotational period relative to the stars; this is the first type of clock, which consequently gradually slows down. The above calculations show the extent of this lag. Astronomical time has long since been found to be slowing down. Even in the eighteenth century, studies of extant material on solar eclipses observed in antiquity showed that these eclipses occurred several hours earlier than what should be expected from calculations based on the theory of motion of the moon and earth. It was at this time, that the German philosopher Kant first suggested the retardation of the earth's rotation.

Careful observations of the moon will reveal a slowing down in the rate of the earth's rotation over a much shorter interval of time. Indeed, in 100 years a uniform clock will advance approximately 18 seconds over an astronomical clock. The moon moves through the stars with a speed of $360°/27.3 \approx 13°.2$ per day or $0".55$ per second. During the added 18 seconds the moon will move something like $10"$. We should thus detect an additional shift of the moon equal to about $10'$ in a century. Such an additional movement (as compared with the findings of lunar theory), which cannot be explained by perturbations of the moon caused by the earth, sun or planets, is actually observed.

But changes in the rate of rotation of the earth are not limited to this so-called *secular retardation*. Occasionally, the terrestrial day experiences oscillations associated with processes that occur within the earth. These oscillations, which in one year attain 0.001 second and more, result in an astronomical clock losing or gaining in this one year from 0.05 to 0.07 second. At the present time, with the aid of a so-called quartz clock—an exceptionally good timekeeper and far more accurate than the rotation of our earth—it is possible to detect directly the nonuniform stroke of the astronomical clock.

19. PROBLEMS OF QUALITATIVE CELESTIAL MECHANICS

To this point we have spoken of studies of the motions of bodies of the solar system which embrace r latively small intervals of time and do not extend too far either into the past or the future. At present, these investigations are successfully carried out with the aid of analytical or numerical methods. Naturally, all these techniques need refining. This is especially true of analytical methods, whose accuracy is far surpassed by that of the numerical methods. We need more precise theories of the motions of satellites, through the use of which we could find out more exactly the compressions of planets, the mass of Saturn's ring, and the masses of the satellites themselves; we need more accurate theories of asteroidal motion, on the agenda is the development of a theory of the rotational motion of the earth to account for varying density of matter in the earth's interior and so forth. However, these are problems that have been studied in quite some detail. It is only a matter of refining and improving our knowledge and the techniques of investigation.

Available analytical and numerical theories of the motions of planets, satellites, asteroids and comets are, in the majority of cases, sufficient to give an accurate description of how these bodies moved tens and even hundreds of years ago and how they will move as many years hence. Yet when studying the problem of the origin and development of our solar system we need to know the nature of motions for much greater periods of time. We are interested in the motions of planets, satellites, asteroids, and comets that occurred hundreds, thousands and millions of years ago, and the changes that have taken place during these periods of time. Of no less importance is the study of changes that can occur in planetary motions millions of years hence.

In all these cases the available theories of motion of these bodies are of no help since they can be applied only for very limited and relatively small intervals of time. For this reason, so-called *qualitative methods* of celestial mechanics are invoked for investigating changes in the character of the motions of bodies over long periods of time. They differ from quantitative methods in that they do not perm t direct computation of the positions of heavenly bodies in space or determination of their mass, etc., but they do allow one to assess changes of motions of a general nature.

Let us examine, for example, the two-body problem in which two bodies are receding from one another at an initial instant. The problem is to find out whether these bodies can recede to any distance (or, technically speaking, to *infinity*) or not.

In this case, it is a simple matter to answer this question without applying qualitative methods because we have a precise solution to this problem. Depending on the initial velocities, the two bodies will move one relative to the other in an ellipse, a parabola or a hyperbola. Therefore, if the velocities of the bodies at the initial instant are greater than a definite magnitude (hyperbolic motion), they will recede from each other infinitely; if less (elliptical motion), they will not be able to separate beyond a specific distance.

Or take a similar çase in the motions of three bodies. Let three bodies of certain specific masses move, at an initial instant, in different directions from one another (that is, the distance between them will at first be constantly incre sing). What will be the future motions of these bodies?

In this case we do not have a precise solution to the problem, but an answer may be given by investigating the problem with qualitative methods.

If the velocities of these bodies at the initial moment are greater than a definite magnitude, all the bodies will recede from each other to infinity. However, if the velocities are less than this value, two cases are possible: either all three bodies will move without receding from each other to more than a specific distance, or one of the bodies will go to infinity while the other two will move in ellipses relative to each other.

Let us take another problem — the motion of two bodies whose masses are not constant but diminish with time. If the law of variation of mass of these bodies is not known we shall not be able to define their motions accurately. But qualitative methods permit saying that if the velocities of the bodies at the initial instant are relatively small, they will move, relative to each other, in ellipses, whose semi-major axes and eccentricities will gradually increase; thus, these bodies will gradually recede from each other and their paths will become more and more elongated.

In the nineteenth century, a great deal of attention was devoted to the problem of the stability of the solar system. This problem may be stated as follows: will all the planets always be moving in almost exact circles in one plane and at nearly the same mean distance from the sun as they are now, that is to say, are variations in the semi-major axes, eccentricities, and inclinations of the planetary orbits purely secular variations?

Extensive investigations of secular perturbations of the planets, which have already been spoken of above, were carried out by Lagrange, Laplace, Leverrier, and others. Their studies demonstrated that there are no purely secular perturbations in the above-mentioned orbital elements of the planets of the solar system. But in these investigations, due to tremendous difficulties of a mathematical nature, only the principal mutual perturbations of the planets were taken into account, while the more insignificant perturbations were disregarded. For this reason, there is no, mathematically speaking, rigorous solution to this problem. The results obtained so far only permit us to state that the planetary motions will be stable for several millions of years.

Certain conclusions about the distant past of the solar system may be drawn from geological investigations. Geological findings indicate that in the course of several hundreds of thousands and even millions of years there did not occur any radical changes in the earth's climate that could be attributed to a change in the nature of the earth's motion about the sun. Even such perceptible changes in the climatic conditions of Europe in the past as the onset of the ice ages may be adequately explained by slight oscillations in the eccentricity of the earth's orbit and in the inclination of the rotation axis of the earth to the plane of its orbit due to long-period perturbations. Investigations show that these oscillations are fully capable of leading to a fall in the mean annual temperature of Europe sufficient to give rise to glaciers. More substantial alterations of the eccentricity of the orbit or of the mean earth-sun distance resulting either in the earth's appreciably approaching the sun or receding from it, would have affected the climate to a far greater extent, but geology is ignorant of even the slightest traces of such radical alterations in the climate over a number of millions of years.*

Hence, we may conclude that the earth has been moving during this time just about as it is at present. And since the motions of all the planets are interrelated due to mutual perturbations, considerable changes in the motion of one planet cannot but lead, ultimately, to appreciable alterations in the motions of all the planets. But since there are no significant perturbations in the motion of one planet, the motions of the others should, apparently, have changed but slightly. From this reasoning it may be inferred that not only the earth's motion but that of the other major planets has been materially the same for millions of years.

A very interesting case is the history of the investigations of Saturn's ring. When viewed even in the most powerful telescope it has the shape of a solid body. From observational data the thickness of the ring is estimated at about 20 km., while the width comes out to roughly 60,000 km. In the eighteenth and nineteenth centuries conjectures were advanced that Saturn's ring is indeed a solid structure. True, some astronomers even at that time were doubtful about the existence around the planet of a solid, very thin and extensive circular plate. The final answer was given in the mid-nineteenth century by the English physicist Maxwell. He reasoned as follows. Since the ring is a material body it must obey the law of gravitation. If it were not in motion it would have to fall onto the planet due to gravity. Therefore the ring is in motion. But can a solid flat thin ring revolve about a planet under the influence of the force with which the planet attracts each particle of the ring? Maxwell demonstrated that such rotation of a solid ring could not be stable. At some time following the beginning of motion the solid ring would have to break up into separate tiny pieces. For this reason, Maxwell came to the conclusion that the rings of Saturn should consist of a number-

---

* Closer to the truth, incidentally, is the suggestion that the ice ages and other large-scale alterations in the earth's climate may be explained by purely "local," or "terrestrial" causes, say, changes in the nature of heavy and permanent ocean currents.

less multitude of tiny solid bodies, each of which has its own orbit about Saturn due to the action of Newtonian gravitation.

Later observations, carried out at the end of the nineteenth and the beginning of the twentieth centuries by the Russian astronomer Belopolsky and the American astronomer Keeler fully corroborated Maxwell's conclusion. They found that the outer parts of the rings of Saturn revolve more slowly than the inner parts. A solid body cannot rotate in such fashion. This means that Saturn's rings are not solid but consist of numerous independent bodies that are in revolution about the planet obeying Kepler's Third Law: the closer to the planet the body, the faster it moves.

There are very many interesting problems for qualitative investigations of the motions of satellites and asteroids associated with problems of the origin of satellites, asteroids and the entire solar system.

For example, if we knew what changes have occurred in the motions of asteroids since very remote times we could get some insight into the conditions under which the asteroids originated. By way of illustration take the following interesting problem in asteroidal studies.

In the case of 98 per cent of the known asteroids, which is more than one thousand five hundred, the mean solar distances (the semi-major axes of the orbits) lie between 2.2 A.U. (330 million km.) and 3.6 A.U. (540 million km.), with periods ranging from 3.2 to 7.0 years. However, these mean distances are not uniformly distributed. For instance, there are hardly any asteroids with mean solar distances close to 3.27, 2.84 and 2.5 astronomical units. Bodies with such mean solar distances should have orbital periods equal to about 5.9, 4.8 and 4.0 years, respectively, which amounts to 1/2, 2/5 and 1/3 of the Jovian period.

This means that asteroids appear to shun orbital periods that are commensurable with Jupiter's period. This gives rise to so-called "gaps" in the distribution of the mean solar distances of the asteroids.

Yet there are two very interesting groups of asteroids somewhat farther from the sun than the main swarm. These groups have orbital periods about the sun that are commensurable with Jupiter's period of revolution. One of these groups consists of 16 presently known asteroids that have a mean solar distance of about 4.0 A. U. and periods of revolution close to eight years. This is the "Hilda Group," named after one of its members. The second group comprises 14 known asteroids and goes by the name "Trojan Group"; these asteroids received the names of heroes of the Trojan war described in Homer's *Iliad* (Achilles, Odysseus, Hector, and others). They have a mean solar distance roughly that of Jupiter (about 5 astronomical units) and, consequently, have approximately the same orbital period (close to 12 years).

A natural question is why are there "gaps" for asteroids near 3.27, 2.8 and 2.5 astronomical units? Why do the asteroids avoid orbits with these mean solar distances? On the other hand, why are orbital periods of 8 and 12 years, which are commensurable with Jupiter's period of revolution, so "convenient" for asteroids that the latter have evolved two groups with these periods?

There are two possible answers:

a) these peculiarities of motion are connected with the conditions of origin of the asteroids; asteroidal motions possessed such peculiarities from the very moment of their "birth";

b) these peculiarities of motion are independent of the original conditions of the asteroids and evolved under the influence of the disturbing forces of the planets.

To confirm one or the other of these points of view, one has to investigate changes in asteroidal motions that have occurred since the very remote past. For instance, if it were possible to demonstrate that asteroids with orbital periods of 5.9, 4.8 and 4.0 years should experience secular perturbations, which in the course of millennia would gradually alter these periods, this would confirm the second view. It would then even be possible to give a rough appraisal of the time required to form the above-mentioned gaps, and thereby also the age of the asteroidal system. Conversely, if it were shown that there are no purely secular variations in the orbital periods of these asteroids, and therefore, that no gaps could have appeared in the process of changing asteroidal motions, we could then say that the gaps had arisen as the asteroids evolved.

This question has already been investigated and the results show that, apparently, the gaps are associated with the perturbing influence of Jupiter. If the orbital periods of Jupiter and an asteroid are commensurable, perturbations of the latter should be much greater than if commensurability were not the case (recall the instance analysed above of the mutual perturbations of Jupiter and Saturn, whose orbital periods are also commensurable). However, this is not, as yet, an exhaustive answer to the problem of the origin of the gaps.

More involved and, so far, less successful are studies of the secular perturbations of the two earlier mentioned groups of asteroids. Peculiar in this respect are the Trojans. These asteroids are isolated from the main swarm and should present an interesting picture. What movements did they have millions of years ago? Could they have been dissociated and then have formed into a group through variations in their motions? Will this group remain intact in the future or will secular perturbations alter the orbital periods of the Trojans and their mean solar distances and disperse the members? Answers to these questions would give us some idea about the possible ages of the Trojans and would help to clarify the conditions in which they originated.

The problem of the formation of groups of asteroids with similar orbits is particularly intriguing since there are a large number of such groups. There is even a hypothesis that all the asteroids originated in the disintegration of a single major planet while the various individual groups of asteroids appeared as a result of repeated disintegration of the larger "fragments" of the splintered planet.

On the one hand, it should be interesting to find out whether such groups of asteroids could have formed as a result of planetary perturbations, and, on the other hand, can secular perturbations produced by planets lead gradually to a dissociation of such groups, if they originated in the process of the formation of the asteroidal system. Such investigations could to some extent confirm or reject the hypothesis at hand.

Unfortunately, in studies of this nature we encounter formidable mathematical difficulties that have yet to be overcome. There are still far more problems concerned with qualitative investigations of the motions of heavenly bodies than there are answers to them. And what is more, these interesting problems, which are so intimately bound up with cosmogony—the science of the origin and development of heavenly bodies—have so far engaged but little attention. Only the future holds a more complete solution to these problems.

Exciting problems arise also in studies of our satellite, the moon, and of the Martian satellites. But in these cases, account must be taken not only of the action of Newtonian attraction but also of tidal friction.

We have already spoken of the forces of tidal friction as gradually retarding the earth's rotation. In addition to this retardation of the rate of rotation, there should also occur a slow increase in the mean moon-earth distance, which means that at an earlier time the moon was closer to the earth than at present. If the earth's period of rotation increases roughly 0.001 second per hundred years, the mean lunar distance should now be increasing at the rate of 2 metres every hundred years. Of course, this is an insignificant figure, but if one considers the motions of the earth and moon over the course of thousands and millions of years, the change effected by tidal friction becomes very perceptible indeed.

Judging from calculations made by the English scientist George Darwin, approximately 4,000 million years ago the moon was at a distance of only 14,000 kilometres from the earth, and the terrestrial day was only 5 hours long. At the beginning of the twentieth century George Darwin even suggested that the earth and moon were one, and that the moon at some later date separated from the earth. In the future, according to this hypothesis, the moon should continue to recede from the earth and the terrestrial day will continue to lengthen. Many thousands of millions of years hence the moon-earth distance will increase by 50 per cent, the lunar period of revolution (the month) will increase to 47 days (628 hours) and the terrestrial day will be of the same length. Darwin then says that at this stage tidal friction should produce just the opposite effect on the moon's motion, and the latter will begin to approach the earth. Ultimately, the moon can approach the earth to such a distance that it will be broken to pieces by the earth's gravitational pull, which will give rise to huge tides on the moon. During this final stage, the moon's orbital period (month) will remain shorter than the earth's period of rotation (day).

Unfortunately, it is difficult at present to say just how true this hypothesis is As to the past, it is very doubtful whether the moon could ever have separated from the earth. Investigations carried out by the noted Russian mathematician Lyapunov showed that a separation of the moon from the earth is impossible. More, when Darwin studied the future evolution of the moon he did not take into consideration any forces, other than gravitation, that could affect the motions of the earth and moon. However, investigations covering large periods of time have to take account of possible changes in the physical structure of the earth and moon and of the entire solar system. But even so, we must admit that tidal friction can be a serious factor influencing the motions of bodies over very lengthy intervals of time.

A hypothesis concerning the effect of tidal friction on the motions of satellites might be an aid in the study of the Martian moons. These satellites are among the most remarkable objects in the solar system. First of all, they are very close to their primary: the first, Phobos, is distant 9,380 km., from the centre of the planet which is a mere 1.5 Martian diameters and only 5,930 km. from the planet's surface, the second, Deimos, is 23,500 km. from Mars, or 3.5 diameters of the planet. Another interesting thing is that Phobos makes a complete circuit of Mars in less than one third of a Martian day (the planet has a period of rotation equal to about 24 hours 37 minutes, while Phobos has an orbital period of roughly 7 hours 39 minutes). In other words, the month on Mars (judging from Phobos) is shorter than the day. If the earth had a similar satellite it would move from west to east instead of from east to west, and would rise in the west and set in the east, in complete contrast to all the other heavenly bodies including our artificial earth satellites.

On the Darwin theory, the orbital period of the satellite (the month) is first longer than the planet's period of rotation (the day), and the satellite gradually moves away from its primary. Then, at a certain stage, having receded to a definite distance, the satellite begins gradually to approach the planet. During this stage the length of the month is less than that of the day. And the satellite can make an extremely close approach to the planet.

The first impression is that Phobos is in the latter stage of this evolutionary scheme: the month on Mars (for Phobos) is shorter than the day, and Phobos is very close to its primary. According to this scheme, the satellite will continue to approach Mars.

On the other hand, Deimos, according to Darwin, should be in the first stage of recession from the planet, since the month (for Deimos) is equal to 30 hours and 18 minutes, which is longer than the day.

Observations conducted between 1879 and 1941 (the Martian satellites were discovered in 1877) confirm the fact that the orbital period of Phobos is gradually diminishing; and at an average of 0.00025 second per year, according to the American astronomer B. Sharpless. This figure corresponds to 646-centimetre per year decrease in the semi-major axis (six metres a century). It follows that within 100,000,000 years Phobos will fall onto Mars (the decrease in 'a' will exceed the present mean distance of Phobos from the surface of its primary).

A complicating factor is that the cause of such a considerable secular perturbation in the orbit of Phobos, in large measure, is still a mystery. The British astronomer H. Jeffreys found (after some theorizing on the internal structure of Mars) that tidal friction was responsible for only 1/1000 of the observed decrease in the semi-major axis. Other astronomers have pointed to the decelerating action of interplanetary matter. This is analogous to the effect of the terrestrial atmosphere on artificial earth satellites. At the same time, the Soviet astronomer N. Parysky has found that tidal friction — notwithstanding Jeffreys' findings — *can* produce the observed effects, if we asume that Mars and the earth have the same viscosity.

The problem of the motion of Mars' satellites remains "on the agenda" as one of the most interesting problems of qualitative celestial mechanics.

20. STELLAR MOTIONS AND THE LAW OF GRAVITATION

The arrangement of stars in the sky remains the same from day to day, and even during a whole lifetime it is impossible to notice any changes in their positions. This is why the ancients called them "fixed stars." However, it is hard to find a name less appropriate. Careful measurements of stellar positions made with powerful instruments show that all the stars are in motion in the sky. In part, these motions are due to the fact that we observe the stars from a moving earth. But, on the other hand, these motions are the actual motions of the stars and also that of the sun carrying with it in space its whole family of planets. Shifts in the apparent positions of stars produced by the motion of the star itself and of the solar system in space are termed *proper motions*. Stars have very small proper motions. The majority move only $0''.01$ per year, which is a small angle indeed—that subtended by a hair at a distance of 2 kilometres! Only a few stars have a proper motion of $1''$ per year, and of these only about ten of the very "fastest" stars move several seconds of arc a year (from 4 to $10''$). This is why naked-eye observations (whose accuracy does not exceed $2'$) do not exhibit any movements in the great majority of stars even during 100 to 200 years. True, if a comparison is made of the positions of stars over many hundreds of years, their proper motions may be detected with relative ease. It was precisely in this way that stellar motions were discovered in ancient China, as recorded by the Chinese chronicles. But Europe did not know of this remarkable discovery of the Chinese astronomers.

In 1718, Halley noticed that three stars (Sirius, Procyon, and Arcturus) had altered their positions as compared with the observations of the ancient Greek astronomers.

Comparing the observed positions of these stars with the data of the Greek astronomer Hipparchus (first century B. C.), Halley found that Sirius had moved $1°.7$, Procyon $0°.7$ and Arcturus $1°.1$. This is how much these stars moved in 18 centuries.

At the end of the eighteenth century, orbital motion was discovered in *binary star* systems.

In naked-eye observations, binary stars frequently do not differ from the other stars. But through a telescope one finds that each binary star consists of two separate stars close together (at a distance not exceeding several seconds of arc). The first fundamental studies of binary stars were carried out by Herschel at the end of the eighteenth and the beginning of the nineteenth centuries. He found that the component stars of a binary system are in motion relative to each other. Herschel discovered several hundred binary stars, at the present time some 20,000 such systems have been recorded.

What forces govern the motions of stars and force them to move about each other?

Stars are balls of incandescent gases just like our sun. They are material bodies that should attract each other according to Newtonian law. The fact that the motions of stars do indeed obey the law of gravitation is above all vividly demonstrated by observations of binary stars.

One star in the pair is usually brighter than the other. This one is called the *brighter component*, or *primary*, the other one the *fainter component* or *satellite*. Careful measurements of the mutual positions of the pair of stars show that the satellite describes an ellipse about the brighter component. And this motion always obeys Kepler's Second Law, the Law of Areas. Which means that the stars move under the influence of the force of mutual attraction. But maybe this force does not obey Newton's law of gravitation, that is, the inverse-square law? No, in the mid-nineteenth century three French mathematicians Alphand, Darboux and Bertrand gave rigorous proof that stars should attract each other in strict accord with Newtonian law. They demonstrated that the motion of the fainter component about the brighter one in an ellipse could occur in two cases only:

1) if the force of attraction increases in proportion to the distance; the brighter component should then be at the centre of the ellipse described by the satellite.

2) if the force of attraction varies in accordance with Newton's law (inversely as the square of the distance); in this case, the brighter component is located in a focus of the elliptical orbit of the satellite.

In all other cases of the dependence of the force of attraction on distance the satellite will not pursue an elliptical path about the main star.

Observations show that the brighter component is never in the centre of the ellipse described by the fainter component. This precludes the first case. What is more, this case is hardly possible physically speaking: one finds it difficult to imagine that the force of attraction of a star *increases* with the distance from this star. We therefore find the attraction in accord with Newtonian law, and the two stars of the pair should move in accordance with the problem of two bodies attracting each other. A brilliant confirmation of the application of Newton's law was the discovery of a companion of Sirius. In 1844, Bessel noted that Sirius was describing a wavy trajectory in the sky. He therefore concluded that Sirius should have an invisible companion and that these two stars should move under the influence of mutual attraction in ellipses about their common centre of gravity. Bessel determined the orbit of the invisible satellite and even estimated its mass. In 1862, after Bessel's death, this companion star of Sirius, whose existence had been predicted twenty years earlier, was found. Its orbit is roughly that which Bessel himself had computed.

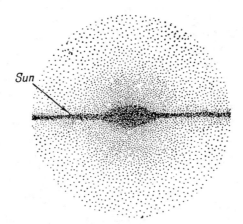

Fig. 50. A schematic view of the Galaxy "on edge."

In addition to binary stars, we also find systems comprising three, four and more stars each. These are called *multiple stars* or *multiple systems.* The stars in a multiple system are located close to each other and move under the action of mutual attraction. The motions of multiple-system components are very intricate since it is no longer a problem of two bodies but of several bodies. Actually, the theoretical study of the motions of such systems has only just begun.

Nearly all the stars that we see in the sky with the naked eye are part of a huge stellar system known as the *Galaxy.*

The Galaxy contains something like 150,000 million stars, the bulk of which form a lens shaped structure when viewed on edge (Fig. 50). The Galaxy is 85,000 light years in diameter, while the densest part near the centre has a thickness of 1,000 light years.*

Thus, the stars of the Galaxy are concentrated mainly near a certain common plane called the *galactic plane*. The sun and its system of planets is situated in the main mass of stars near the galactic plane at a distance of some 23,000 light years from the centre of the Galaxy. Since the solar system is situated near the galactic plane, we see the major portion of the stars in a direction parallel to this plane. The luminous patch of the Milky Way that girdles the sky is the main accumulation of stars of the Galaxy. When looking perpendicular to the galactic plane we see, on the average, far fewer stars.

How do the stars in the Galaxy move?

The observed proper motions of the stars in the sky appear, at first glance, to be without any system whatsoever. But a careful and long term study of the motions of stars has shown that this is not the case at all. It turns out that on the average (that is, if we disregard the individual peculiarities in the motions of the separate stars, or clusters) all the stars are moving around the centre of the Galaxy in one and the same direction, so that the Galaxy appears to be rotating about its centre. Of course, this is not rotation in the proper sense of the word. Each star has independent motion and its velocity depends on its distance from the centre—the farther away it is the slower it moves on the average—and it is only due to the fact that stellar motions occur mainly in a single direction that these motions collectively give the impression of rotation.

The forces which govern the motions of stars and make them revolve about the centre of the Galaxy are the Newtonian forces of mutual attraction. Each individual star is acted on by the total attraction of the millions and millions of other stars of the Galaxy. Since the stars of the Galaxy form a system that is symmetric with respect to the centre

---

* A light year is the distance light travels in one year, or 9.5×10^{12} km. Light covers the distance from sun to earth in only 8 minutes and 18 seconds!

this integrated attraction is directed to the centre of the Galaxy and each star is attracted to this centre. However, the law of the variation of this attraction with distance from the centre and distance from the galactic plane is extremely involved and far from being understood. In this case, naturally, there can be no talk of gravitation in strict accord with the Newtonian law, that is, strictly in inverse proportion to the square of the distance from the centre.

The problem of the galactic law of attraction may be approached from two angles. The first is a theoretical approach. It amounts to calculating theoretically what the gravitational pull will be on a single star in the Galaxy if each of the 150,000 million stars occupying a definite volume of space attract this star in accordance with Newton's law. In addition to the purely mathematical difficulties that arise, we have also to deal with fundamental difficulties since we do not know exactly how these 150,000 million stars are arranged in space. And without this knowledge it is impossible to give an accurate calculation of the force of attraction of the stars. This can only be done approximately, on the assumption, for example, that the stars in the Galaxy are distributed uniformly (this is the simplest assumption), or assuming that the density distribution of stars varies with the distance from the centre of the Galaxy in accordance with some definite law.

The second approach to the problem may be called an empirical approach. In this case, the observed motions of stars are used to determine the forces which should bring about these motions (the problem of determining a force from a given motion). This was the approach made by the Soviet astronomer P. Parenago, who showed that the stars in the galactic plane are attracted towards the centre with a force

$$F \sim \frac{R}{(1 + aR^2)^2}$$

R being the distance from the centre of the Galaxy, and a a certain constant. The expression for the law of attraction of stars outside the galactic plane is far more complicated.

What is needed, of course, is investigations that would unite these theoretical and empirical methods and utilize both the diversity of observational findings and the theoretical attainments of celestial mechanics.

In bringing to a close our discussion of the motions of stars we can say that gravitation governs not only the motions of the planets of the solar system but also those of the most remote stars and stellar systems.

21. WHAT IS GRAVITATION

The problem of the nature of the forces of gravitation arose a long time ago, immediately after Newton discovered the law of universal gravitation. What is gravitation, then? What is it that makes material particles attract each other? These are the questions that immediately arise when one learns of the Newtonian law. But it is no easy matter to answer such queries. Physically, gravitation is very difficult to understand. This was the reason why Newton's theory was at first met with incredulity. Many scientists tried to disprove it and even denied the existence of gravitation. In time these doubts disappeared and the validity of Newton's law became generally recognized. But the nature of the gravitational force remained incomprehensible, even mysterious.

Wherein lies the complexity of this problem?

First of all, gravitation manifests itself as action *at a distance*. Indeed, the force of gravitation acts between bodies no matter how far away from each other they are; and it is unimportant that airless space separates them. Newton's contemporaries thought that this ran counter to direct experiments which indicated that bodies act on one another only through contact (the transmission of motion by pushing, pressure, thrust, etc.). Newton himself expressed the view that gravitation is transmitted from one body to another via the "ether"—a peculiar tenuous medium that pervades the entire space between bodies and fills the interstices in all solid bodies. Some scientists after Newton attempted to give a more detailed explanation of gravitation through the action of the ether.

An opinion was also voiced that gravitation is the result of mechanical action on bodies by invisible "ultra particles," moving in all directions in space. If a certain body were alone in space it would receive impacts from these "ultra particles" uniformly from all sides and would therefore be in a state of equilibrium. But if, for example, there were two bodies they would shield each other from the impinging particles moving exactly along the line connecting the centres of the bodies, and the particle fluxes would push the bodies towards each other, thus creating a force of attraction between them.

In certain hypotheses, the action of "ultra-particles" is replaced by a similar action of streams of ether.

The originators of such hypotheses strove to eliminate the mysterious action at a distance of gravitational forces. Yet they were unable to explain other strange properties of gravitation.

First among them was the apparent instantaneousness with which gravitation acted.

Indeed, when applying Newton s law it is taken that the forces of attraction depend only on the mutual positions of the bodies and that as soon as they change, the magnitudes of these forces change instantaneously, which means that gravitation is transmitted instantaneously. Now if the action is transmitted through the agency of some medium (ether) we need to find out the rate at which this action propagates. Similarly, the rate of propagation of gravitation has to be determined if the latter is thought to be due to moving "ultra-particles" or streams of ether. And if such be the case, we must introduce corrections for this velocity when computing the gravitation between bodies, for if some body A arrives at a point located at a distance r from body B at the instant t_0, then at this instant, A will not be able to attract B with a force proportional to $1/r^2$ on the condition that gravitation propagates with a definite velocity. A certain time would be required for gravitation to be transmitted from body A to body B. But during this time A would reach another point of space.

Thus, at each instant the force would depend not only on the configuration of bodies but also on their velocities and on the rate of propagation of gravitation. In the absence of such corrections we should observe deviations in the motions of bodies from those computed on the basis of Newton's law. Yet no such deviations are observed. This could occur if the velocity of gravitation is so great that it has practically no effect on motion, which is to say that in practice we have to do with instantaneous propagation of gravitation. Laplace calculated the minimum velocity of gravitation that would render corrections for this speed undetectable in observations of astronomical objects (the moon, for instance). The result was a magnitude at least a million times the velocity of light. At the present time, the existence of such a speed of motion of particles or propagation of physical phenomena in a medium is believed to be physically impossible.

Second, gravitation is a force that recognizes no barrier. It is neither depleted nor absorbed either by an interstellar medium or when it encounters bodies. For instance, during lunar eclipses the earth passes between the sun and moon and thus could block the forces of gravitation between them exactly as it does rays of light. It is possible to determine the perturbations in lunar motion that should arise if the solar pull on the moon slackened during eclipses. But no such perturbations are observed.

No known material used as a shield or screen has been even partially effective in stopping the force of gravity, which attracts all material bodies to the centre of the earth.

How is one to picture particles, or a tenuous medium, capable of propagating instantaneously and penetrating any bodies unchecked without alteration?

Nowhere in nature do we encounter such particles, fluxes of particles or such a manner of propagation of physical phenomena and processes. Take light, X-rays, radio waves, electric and magnetic forces, they are all absorbed to a greater or lesser extent by material bodies, a material medium, and all of them propagate with a finite speed (equal to the velocity of light). Thus, a primitive mechanical interpretation of gravitation based on the simple mechanical action of an ether or of fluxes of ether, or fluxes of particles is unable to explain its amazing properties. Today this interpretation of gravitation appears naive in the extreme and runs counter to present-day physical concepts. But how is one to explain this "marvellous" all-permeating force of gravitation that acts instantaneously at a distance?

The answer was given by one of the greatest of modern scientists, Albert Einstein (1879-1955). Einstein is the author of a large number of outstanding studies in many fields of physics, but his greatest achievement was the *special theory of relativity*, which might be called a "theory of space and time," and the *general theory of relativity*, or, more appropriately, a "theory of gravitation." These theories have wrought great changes in our notions of space and time, about the motions of material bodies and concerning the interrelationship of space, time, and gravitation. The special theory of relativity has demonstrated that space and time are intimately related and that the mass of any body is connected with its energy. The general theory of relativity shows that the spatio-temporal properties of the surrounding world (its *geometry*) are determined by material bodies.

Unfortunately, both the special and general theories of relativity are very difficult to explain in popular language. On the one hand, this is because of the difficulty of the theory itself, of the complexity of its mathematical apparatus, and on the other hand, because of certain unexpected conclusions that follow from the theory and are frequently a radical departure from our "common sense" notions. Nevertheless, we shall attempt to give the reader some idea of what the theory of relativity is about.

In *classical mechanics*, the mechanics of Galileo and Newton, there is no interconnection between space, time and material bodies in space. The diverse processes that occur in space are measured by a time that flows uniformly and is independent of space and independent of the material bodies in space. The properties of space—whose sole purpose is to be a passive receptacle for material bodies—remain always constant and immutable, independent of the distribution of material bodies and even irrespective of whether there are any material bodies present or not. The geometry of this space is Euclidean, the geometry which all of us studied at school. It teaches that the shortest distance between any two points in space is a straight line; it is precisely this straight line that a ray of light follows when moving from one point in space to another. In classical mechanics the mass of a body is constant and unchangeable, as also are the geometrical properties of a body (length, width, shape, etc.) which remain the same irrespective of whether the body is in motion or at rest.

However, such notions, though perfectly natural and commonplace to all of us and apparently justified by our daily practice, are only an approximate reflection of reality. Nature, it appears, has established a far more profound interrelationship between space, time, and matter. Euclidean geometry no longer holds in a space with material bodies. The masses of bodies and their geometry depend on their velocities of motion. The properties of motion of material bodies are considered to be the result of the geometrical properties of the space and not of the forces of attraction.

Recall the law of inertia. In the absence of a disturbing force a body will remain at rest or in a state of uniform motion in a straight line. Let us try to verify this law by experiment. We know that the principal reason why a moving body, in terrestrial conditions, stops after the force ceases to act is friction. Let us suppose that there is no friction. Will the body remain in uniform motion and in a straight line? The answer is no, since the body will be acted upon by the gravitational force of the earth, moon, sun and planets. We go further and remove the entire solar system, the closest stars, even all the stars in the Galaxy. We may then contend, with a high degree of accuracy, that the body will remain in uniform motion and in a straight line. But if there is a star (or any material body, for that matter) somewhere near our body, the latter will have a path that is no longer rectilinear; due to the gravitation of the star the path will be curved.

Now let us compare two cases. In the first case, a single, isolated body is in motion in space in a *straight line*. Insofar as this body is moving in "empty" space, we may say that the geometrical properties of empty space are such that the path of a body in this space is a *straight line* no matter in which direction it is moving. In the second case, the body is not moving in empty space but in the field of gravitation of another material body. In this case, the path of our body is a *curved line*. Which means that the presence of material bodies alters the geometrical properties of space: in the absence of material bodies space is homogeneous, which fact finds expression in the total equality of directions in which an isolated material body can move by inertia. In the presence of material bodies, space becomes curved.* It is this inhomo-

---

* By "curvature" of space we mean that the shortest distance between two points in "curved" space is not a straight line but a *curved* line. By way of illustration, let us take a plane and a sphere. We shall measure the distance between two points of the sphere and plane, so as not to go beyond the limits of the plane or sphere. To do this, one

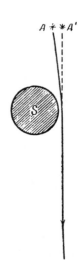

$A + *A'$

Fig. 51. Deflection of a ray of light near the sun. The ray of light from star A is bent, and the star appears to be displaced from the sun (to position A')

geneity of space, its "curvature" that we perceive as gravitation.

Thus, according to Einstein's theory, gravitation is a manifestation of the space-time properties of the world; the distribution of material bodies determines the *geometry* of the space in which these bodies are located and also the motions of these bodies.

Proceeding from these conceptions, Einstein developed his theory mathematically and obtained a series of remarkable results that have been fully confirmed by experiment.

In his works, Einstein resolved the mystery of the enormous rate of propagation of gravitation and its amazing penetrating ability. But, unfortunately, these problems require so complex a mathematical apparatus that it is far beyond the scope of this book. We shall therefore deal only with two of the remarkable results of the general theory of relativity obtained by Einstein theoretically and later brilliantly confirmed experimentally.

On Einstein's theory, not only ordinary material bodies, but also light rays are subject to the attraction of material bodies and for this reason should deviate (though very slightly) from their rectilinear path when passing near celestial bodies. The propagation of light is not precisely rectilinear. This sounds very strange at first thought and seems to contradict our everyday experience here on earth. No one had ever observed this to be the case. Yet Einstein proved to be right. During solar eclipses one can observe stars right near the limb of the sun. Accurate measurements of the positions of these stars during eclipses show that they are not in their usual places but slightly displaced from

has only to take a piece of thread, connect the two points and measure the length of the thread. On the plane, it will lie on a straight line, while on the sphere the line will be curved—the arc of a large circle on this sphere.

the sun (Fig. 51). This displacement is small (roughly 2″) but it is exactly as predicted by Einstein's theory. Einstein's theory of gravitation is also confirmed by certain peculiarities in the motion of Mercury. As far back as the middle of the nineteenth century, Leverrier noted that Mercury's observed position was slightly different from that predicted by Newtonian theory with account taken of the perturbations produced by all planets. This discrepancy had to do with the secular motion of the line of apsides of Mercury's orbit, and it was exceedingly small: Mercury's perihelion was moving 43″ faster per century than suggested by classical theory. This problem was the subject of many studies in celestial mechanics. Attempts were made to explain this divergence by the fact that Newton's law was not altogether exact, that the force of gravitation did not vary exactly in inverse proportion to the square of the distance. But if this change helped to harmonize calculations and observations in the case of Mercury, it yielded divergences between theory and observations with respect to the other planets. And other attempts were made to eliminate this discrepancy, but not one of them produced the desired result.

Einstein's theory of gravitation gave the clue to the mysterious motion of Mercury. This theory says that in a two-body problem (the sun and a planet) the orbit of the planet is an ellipse which slowly turns in space. The rate of this motion depends on the mass of the sun and the distance of the planet. For Mercury, the line of apsides should turn 43″ in one century, which is exactly the amount that was lacking in the earlier theories of motion of Mercury. Thus, the former riddle became a remarkable confirmation both of the accuracy of the analytical theories of the major planets and of the correctness of Einstein's theory of gravitation.

Though Einstein's theory of relativity changes our everyday views concerning space, time, the motions of bodies and their interactions, and, as far as the fundamentals go, departs radically from the classical mechanics and gravitational theory of Newton, its practical application produces nearly the same result in the majority of cases. But in what cases? Only when the velocity of motion of the body is relatively small in comparison with that of light (300,000 km/sec.), and also when we have to do with comparatively small masses. If we disregard the velocities of motion of the bodies v as compared to the speed of light, that is, if we take the ratio v/c to be zero, then all of Einstein's equations and ratios convert to the ordinary ones of classical mechanics, in which case all the concepts of classical mechanics, including Newton's law of gravitation, are valid. For this reason, when we consider the motions of bodies under terrestrial conditions, and the motions of planets, asteroids, satellites in the solar system, and stars in the Galaxy, whose velocities are far below that of light, the results obtained by the Newtonian law of gravitation are very close to those based on Einstein's theory of gravitation. But there still are discrepancies. If we have both a precise theory of motion based on Newtonian law and sufficiently accurate observations obtained over a long period of time, these divergences may be detected, as witness the motion of Mercury.

If we consider the motions of material particles due to the force of gravity near the surface of the earth, the various Einstein effects are practically absent due to the earth's comparatively small mass. But if we wanted to study the motion of such a particle on the solar surface, these effects would have to be taken into account in view of the sun's huge mass.

At this point we may ask: is Einstein's theory absolutely correct? Of course not, since no mathematical theory of the phenomena of nature is capable of taking into account and describing the infinite diversity and countless interrelationships of these phenomena. Any concrete scientific theory is a certain approximation to reality, it is a step along the path of man's penetration into the phenomena of nature. Newton's law of gravitation was the first step in the study of the interaction of heavenly bodies. Einstein's theory is a deeper reflection of reality, but it, too, represents but a further step along the endless path leading to truth.

In sections 3 and 4 we applied several theorems without proof. We now examine the proofs of these theorems contained in Newton's *Principia*. In the main, we shall adduce the proofs by the methods that Newton himself used. In this case, elementary mathematics proves to be insufficient, and Newton used his own, new methods that laid the foundation of higher mathematics (the differential and integral calculus). However, Newton clothed these involved methods in a rather simple geometrical form, thereby making the proofs readily comprehensible to readers without a knowledge of higher mathematics in its present form.

Theorem I

The areas which revolving bodies describe by radii drawn to an immovable centre of force lie in the same immovable plane, and are proportional to the times in which they are described.

Proof

Let us designate by S an immovable centre of force, and by P a material body moving under the attraction of this centre (Fig. 52). Let us first assume that the force does not act on P continually but only by momentary impulses that occur in equal intervals of time Δt. Let us designate by t_1, t_2, t_3, etc., the equally spaced instants at which the force of attraction of the centre S acts.

During the first interval of time, from t_1 to t_2, during which the force does not act, the body P will move by inertia uniformly and in a straight line with a velocity v_0 and will cover the distance $P_1\,P_2 = v_0\cdot\,\Delta t$. At the instant t_2, P

will arrive at P_2. If at this point the body P were not acted upon by a force it would continue to move along the line $P_1 P_2$ with the same velocity v_0; at the instant t_3 it would have arrived at point a $(P_2a = P_1 P_2)$. But by definition when body P arrives at point P_2 at instant t_2 it is instantaneously acted upon by an attracting force ("at once with a great impulse," as Newton writes). This force will accelerate the body in the direction $P_2 S$ and will deviate it from rec-

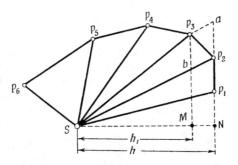

Fig. 52. Proof of Theorem I

tilinear motion along the line P_2a. But since the force acts instantaneously ceasing immediately, the body acquires at once a certain additional velocity v_1 in the direction P_2S, and will then again be in uniform and rectilinear motion (but now with another speed and in another direction). Thus, the velocity of the body, P, during the interval of time from t_2 to t_3 will be compounded of the velocity v_0 in the direction P_2a and the velocity v_1 in the direction P_2S. If the body were moving with a velocity of only v_0, it would cover the distance P_2a during the time Δt; and if it had a velocity of v_1 only, it would cover during this time the distance $P_2b = v_1 \Delta t$. But due to the composition of these velocities the body will move along the diagonal P_2P_3 of a parallelogram constructed on the sides P_2a and P_2b and during the interval of time from t_2 to t_3 it will arrive at point P_3.

Naturally, the diagonal P_2P_3 lies in the same plane as P_2a and P_2b, that is, in the plane of the triangle $SP_1 P_2$ in which the body moved during the interval of time from t_1 to t_2.

Let us denote the altitudes SN and SM of the triangles SP_1P_2 and SbP_3 by h and h_1. Then $MN = h - h_1$ will be the altitude of triangle bP_2P_3. The area of the triangle SP_1P_2 is equal to $\frac{1}{2}h \cdot P_1P_2$. Now the area of the triangle SP_2P_3 is:

$$\text{area } SbP_3 + \text{area } bP_2P_3.$$

However, since

$$bP_3 = P_2a = P_1P_2,$$

the area of the triangle SbP_3 is equal to $\frac{1}{2}h_1 \cdot P_1P_2$, and the area of the triangle bP_2P_3 is equal to $\frac{1}{2}(h - h_1) \cdot P_1P_2$.

Thus, area $SP_2P_3 = \frac{1}{2}h_1 \cdot P_1P_2 + \frac{1}{2}(h - h_1) \cdot P_1P_2 = \frac{1}{2}hP_1P_2$, so that the area of the triangles SP_1P_2 and SP_2P_3 are equal.

By similar reasoning we see that if the attracting force acts instantaneously at points P_3, P_4, P_5 . . . and makes the body move along P_3P_4, $P_4 P_5$ . . ., all these latter sections will lie in a single plane, and the areas of the triangles SP_2P_3, SP_3P_4 . . . will be equal. But the areas of all these triangles are areas described in equal intervals of time by radius vectors from the centre S to the body P.

Now let us take the intervals of time not between adjacent instants, but, for example, between instants t_1 and t_3, t_3 and t_5, etc., which are also equal to each other. During the interval of time $t_1 t_3$ the body P will describe the broken line $P_1 P_2 P_3$, and during the interval of time $t_3 t_5$, the broken line $P_3 P_4 P_5$. The areas of the figures $SP_1P_2P_3$ and $SP_3 P_4 P_5$ will be equal since they consist of equivalent triangles. And, in the general case, no matter what equal intervals of time between any equally spaced moments of action of a force we take, the radius vector of the body P will describe equal areas during these intervals.

Let us now reduce more and more the intervals of time between the impulses of force. The impulse force will then approach more and more a continually acting force. And the lengths P_1P_2, P_2P_3 . . . will also diminish, and the broken line described by the body will differ less and less from a smooth curve. But the deduced property of motion will be retained since the size of the sections did not play any role in the derivation. Summarizing, we can make our impulse force as close to a continually acting force as we like, and our broken line as close to an actual curve described by a body under the influence of the attraction of S, and the property that has been proved will be retained. It will, consequently, hold in the limit, when the intervals, Δt, between the force impulses tend to zero, that is, for the case of a continually acting force. The body P will move along a curve lying in a single plane and concave towards S. In equal intervals of time the radius vector SP should describe equal areas. In other words, given a central force, the areas described by a radius vector are proportional to the times. These areas are equal to the product $k(t'-t'')$. Where k is a constant called the area constant and $t'-t''$ is a corresponding interval of time. Theorem I is thus proven.

Theorem II

If a material body moves in a curved line in a plane so that the radii drawn to a certain immovable point describe equal areas in equal intervals of time, this body is acted upon by a force directed towards this immovable point.

Proof

Figure 53 shows a portion of the path of body P and an immovable point O. On this curve we note positions a_1, a_2, a_3 . . . which P occupies in equally separated instants of time t_1, t_2, t_3 . . . and connect a_1, a_2, a_3 . . . with O, thus dividing the sector Oa_1a_n into a large number of small (elementary) sectors. On the one hand, insofar as the instants of time are equally separated, the areas of any two elementary sectors are equal. On the other hand, insofar as all portions of the arcs a_1a_2, a_2a_3, etc., are very small and differ but very slightly from the sections of straight lines, the areas of these sectors hardly at all differ from the areas of the corresponding triangles $Oa_1a_2, Oa_2a_3, \ldots$

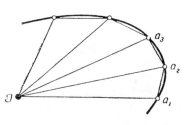

Fig. 53. Proof of Theorem II

In the limit, when the intervals of time between instants t_1, t_2, t_3, ... tend to zero, the ratio: area of elem. sector to area of elem. triangle, is equal to unity. Therefore, in the limit the areas of all elementary triangles are equal.

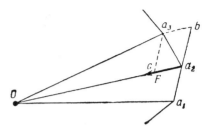

Fig. 54. Proof of Theorem II

Now let us replace motion along our curved line by motion along the broken line $a_1 a_2 a_3$, ..., a_n. But this motion along the broken line is to be regarded as occurring under the action not of a continuous force, but of an impulse force acting at instants $t_1, t_2, t_3, ...$ The subsequent reasoning is just about the same as in the proof of the preceding theorem.

To make the idea clear Fig. 54 gives a magnified section of the path $a_1 a_2 a_3$. During the interval of time $t_1 t_2$ the body P covers $a_1 a_2$ and at instant t_2 reaches position a_2. We extend section $a_1 a_2$ an equal distance to point b. If the body were not acted on by a force it would move further along $a_2 b$ and at instant t_3 would arrive at point b. But at instant t_2 the body is acted on by a force that alters the direction of motion, and the body arrives at point a_3 at instant t_3. The portion $b a_3$ determines the direction of the additional velocity acquired by the body due to the action of the central force S at time t_2, that is, the direction of this force at the given instant. We draw $a_3 c$ parallel to $a_2 b$ and compare the triangles $O a_1 a_2$ and $O a_2 a_3$. Then, doing the same calculations used in the proof of Theorem I it can be demonstrated that the areas of these triangles are equal only when the figure $a_2 b a_3 c$ is a parallelogram. The portion $b a_3$ should therefore be parallel to $a_2 c$.

Thus, the force which attracts P at the instant it is at point a_2 should be directed along the line $a_2 O$. A consideration of the succeeding triangles shows that the force acting at the instant a_3 should be directed along the line $a_3 O$, and so forth.

Now let us pass to the limit by reducing the intervals of time Δt. We see that in the limit: a) motion along the broken line coincides with the actual motion along the curved line; b) the areas of the triangles under consideration are equal and coincide with the areas of the corresponding sectors; c) the force that makes P move is directed towards point O. In other words, body P moves under the action of an attracting force emanating from point O.*

Theorem III

If a body, P, is in motion due to the action of an attracting force of a centre, S, located in one of the foci of an ellipse, the force with which the centre S attracts the body P is inversely proportional to the square of the distance of P from S.

Proof

We shall divide the proof of this theorem into three parts. First we shall derive an important formula that will permit us to relate the force of attraction of an immovable centre, S, and the geometric properties of the curve that the body P describes under the action of this force. After this we shall examine some of the properties of an ellipse and then prove the theorem.

We consider a body P of mass m moving under the action of attraction to a centre S along a certain curve (Fig. 55). Let us assume that the body moves in the direction indicated in Fig. 55 by the arrow, arriving at a certain instant of time t_o at point P_0, and in a very small interval of time Δt at point P_1.** We draw a tangent P_oN to the curve at point P_0. If at the instant the body arrived at point P_0 the attracting force ceased to act, the body P would continue to move with the same velocity (in direction and

---

* Note the essential difference between the broken lines in Figs. 52 and 53. The broken line a_1, a_2, ..., a_n, in Fig. 53 is *inscribed* in the curved line that represents the actual path of the body, while points P_1, P_2, ... in Fig. 52 do not, generally speaking, lie on the curve described by the moving body, so that the broken line P_1, P_2, ... P_5 will neither be inscribed nor circumscribed with respect to this curve. It will only approach it when the intervals of time Δt are reduced and coincide with it in the limit.

** To make the idea more evident, the length of the arc $P_0 P_1$ in Fig. 55 is greatly enlarged.

magnitude) as it had at this
point, that is to say, it would
move along the tangent P_oN
and in an interval of time
Δt would reach point N. On
the other hand, if the body P,
on arriving at point P_0, lost
its velocity entirely, it would
move only in the direction
P_0S due to attraction towards

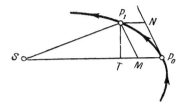

Fig. 55. Proof of Theorem III

S, and during the time interval Δt it would cover the distance P_0M.

Inasmuch as the interval Δt and the P_0P_1 distance covered by P are very small we may regard the force acting during this interval of time on body P as not having changed. Then the acceleration j of the body P, which according to Newton's Second Law is equal to $j=F/m$, will also remain constant. Applying a well-known formula for the path in the case of uniformly accelerated motion, we find:

$$P_0M = {}^1/_2 \; j \; (\Delta t)^2 = {}^1/_2 \frac{F}{m} (\Delta t)^2.$$

Reasoning more rigorously, we must say that when Δt tends to zero the ratio of P_0M to $\frac{1}{2} F/m (\Delta t)^2$ is equal to unity. Since in this ratio the quantities $P_0 M$ and Δt vary while F and m are constant, the limit of the ratio $\frac{P_0M}{(\Delta t)^2}$ is equal to $\frac{1F}{2m}$.

Due to the composition of these two motions, the body P will arrive (within the time Δt) at point P_1, which should be the vertex of a parallelogram constructed on lines P_0M and P_0N. Thus, $NP_1 = P_0M$.

Besides, we know that motion due to a central attracting force obeys the law of areas. For this reason, the area of sector SP_0P_1 is proportional to the time:

$$\text{area } SP_0P_1 = k. \, \Delta t.$$

But if the portion of the arc P_0P_1 is small, the area of sector SP_0P_1, is, in magnitude, close to the area of the triangle SP_0P_1, which is equal to $1/2 \, SP_0. \, P_1T$; these areas coincide in the limit.

Hence it follows that

$$\frac{P_0 M}{(\Delta t)^2} = \frac{k^2 P_0 M}{(\text{area} SP_0 P_1)^2} = K^2 \frac{NP_1}{(^1/_2 SP_0 . P_1 T)^2} \cdot \frac{(^1/_2 SP_0 P_1 T)^2}{(\text{area} SP_0 P_1)^2}.$$

Thus we see that the limit of the ratio $\frac{P_0 M}{(\Delta t)^2}$, equal to $F/2m$, coincides with the limit of the ratio $4k^2 \frac{NP_1}{(SP_0 . P_1 T)^2}$. And the force itself is equal, in magnitude, to the limit of the ratio $8k^2 m \frac{NP_1}{(SP_0 . P_1 T)^2}$.

Utilizing accepted mathematical symbols and noting that only NP_1 and $P_1 T$ vary when P_0 tends to P_1, we may write

$$F = \frac{8k^2 m}{(SP_0)^2} \lim_{P_0 \to P_1} \frac{NP_1}{(P_1 T)^2}. *$$

This limit is obviously dependent only on the geometric properties of the given curve.

This equation enables us to find the magnitude of the force that makes a body move if the path of the body is known.

Now let us assume that the curve along which the body P is moving is a portion of an arc of an ellipse with a focus at S and the centre at O (Fig. 56). We denote the semi-axes of this ellipse by a and b. We draw the diameter of the ellipse $P_0 D$** and continue the straight line $P_1 M$ until it intersects the straight line $P_0 D$ in a point L. Then, on the basis of the properties of an ellipse, we may write the ratio:

$$\frac{P_1 N}{(P_1 T)^2} = \frac{a}{b^2} \cdot \frac{P_0 D}{2DL} \cdot \left(\frac{P_1 L}{P_1 M} \right)^2. \tag{1}$$

This ratio is not at all apparent and is derived as follows.

Through the centre of the ellipse we draw a diameter CG, which is conjugate with the diameter $P_0 D$. A diameter parallel to the tangents drawn through the ends of a given diameter (in this case, through points P_0 and D) is called

---

* The notation $\lim_{x \to x_0} y\,(x) = a$ signifies that "$y(x)$ tends to a limit equal to a when x, approaches x_0."

** The diameter of an ellipse is a straight line connecting two points of the ellipse and passing through its centre.

a diameter *conjugate* with the given diameter. Let us introduce the designations $P_0O = a_1$ $CO = b_1$. We shall first show that if E is the point of intersection of the diameter CG and the straight line P_0S, then the length P_oE is equal to the semi-major axis of the ellipse a.

From the properties of an ellipse we know that a tangent to the el-

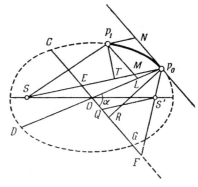

Fig. 56. Proof of Theorem III

lipse forms equal angles with radius vectors drawn from the foci to the point of tangency.

For this reason, if we connect P_0 with the other focus of the ellipse S' and continue this line to an intersection with the continuation of the diameter CG, then $P_oE = P_oF$. We draw $S'Q$ parallel to P_0S. The triangles SOE and $S'OQ$ are equal, and the triangle $S'QF$ is an isosceles triangle. Consequently, $SE = S'Q = S'F$ and $SP_0 + P_0S' = P_0E + P_0F$, i. e.,

$$P_oE = \frac{SP_0 + P_oS'}{2}.$$

But the sum $SP_0 + P_0S'$ is equal to $2a$ for the ellipse; hence, $P_0E = a$.

From the similarity of triangles P_oML and P_0EO it follows that

$$\frac{P_0M}{P_0L} = \frac{P_0E}{P_0O} \ or \ \frac{P_0M}{P_0L} = \frac{a}{a_1} \ or \ \frac{P_1N}{P_0L} = \frac{a}{a_1}. \tag{2}$$

Now, we drop from P_0 a perpendicular P_0R onto the diameter CG, and from P_1 onto SP_0. Since the triangles P_1TM and P_0ER are similar,

$$\frac{P_1T}{P_1M} = \frac{P_0R}{P_0E}.$$

Designating the angle between P_0O and OF by α we find, from the well-known property of conjugated diameters that

$a_1 b_1 \sin \alpha = ab$. And in the triangle P_0OR, the side $P_0R = a_1$ $\sin \alpha$. Therefore, $P_0R = \dfrac{ab}{b_1}$

and, consequently,

$$\frac{P_1T}{P_1M} = \frac{ab}{b_1 a} = \frac{b}{b_1}. \tag{3}$$

According to the properties of conjugated diameters

$$\frac{OL^2}{a_1^2} + \frac{(P_1L)^2}{b_1^2} = 1.$$

From this equality we obtain:

$$(P_1L)^2 = \frac{b_1^2}{a_1^2}(a^2_1 - OL^2) = \frac{b_1^2}{a_1^2}(a_1 + OL)(a_1 - OL).$$

Since $a_1 + OL = DL$, $a_1 - OL = P_0L$,

$$\frac{(P_1L)^2}{P_0L \cdot DL} = \frac{b_1^2}{a_1^2}. \tag{4}$$

Using equalities (2), (3), and (4) as a basis, we obtain from the ratio $\dfrac{P_1N}{(P_1T)^2}$ the following:

$$\frac{P_1N}{(P_1T)^2} = P_0L \frac{a}{a^2} \frac{b^2_1}{b^2(P_1M)} = \frac{aa_1}{b^4} \cdot \frac{1}{DL} \frac{(P_1L)^2}{(P_1M)^2} = \frac{a}{b^2} \frac{P_0D}{2DL} \cdot \frac{(P_1L)^2}{(P_1M)^2}.$$

Thus, we have derived the required ratio (1).

If we now make point P_1 approach P_0, point L will tend to P_0, while point M will tend towards point L. Thus, in the limit, when $P_0 \to P_1$, we obtain

$$\lim_{P_0 \to P_1} \frac{P_1N}{(P_1T)^2} = \frac{a}{2b^2}.$$

Comparing this expression with the earlier obtained expression for the force F, we see that when moving along an ellipse

$$F = m \frac{4k^2a}{b^2} \frac{1}{(SP_0)^2}. \tag{5}$$

Since the quantities k, a, and b are constant when moving along the given ellipse, it follows that the force acting on P varies in inverse proportion to the square of the distance between P and S, which proves the theorem.

Theorem IV

If several material bodies are in motion in ellipses due to the action of an attracting centre of force, S, which varies in inverse proportion to the square of the distance from S, the squares of the orbital periods of these bodies are as the cubes of the semi-major axes of their elliptical orbits.

Proof

The theorem states that a centre of force S attracts any material body P of mass m, at a distance r, with a force.

$$F = m\frac{L}{r^2},$$

L being the constant of proportionality.

If this body is moving in an ellipse with a focus at S, then from equation (5) obtained in the proof of the preceding theorem, it follows that the constant of proportionality L is equal to $4k^2\,a/b^2$, where k is the "area constant" whose magnitude is determined by the rate of motion in the given ellipse, and a and b are the semi-axes of the ellipse. Thus, in the case of motion along different ellipses the area constant and the semi-axes of the ellipse can differ, but the ratio $4k^2a/b^2$ retains a constant value equal to L.

During the time equal to the orbital period T, the body P will complete one full circuit about S and its radius vector will describe an area equal to the area of the entire ellipse, πab, in which P revolves about S.

According to the definition of the area constant k

$$\pi ab = kT \quad \text{or} \quad T/a = \pi b/k.$$

We raise this equality to the second power

$$\frac{T^2}{a^2} = \frac{\pi^2 b^2}{k^2}$$

and rewrite the equality obtained in the following form:

$$\frac{T^2}{a^3} = \frac{4\pi^2 b^2}{4k^2 a}$$

or

$$\frac{T^2}{a^3} = \frac{4\pi^2}{L}.$$

Summarizing, the ratio T^2/a^3 remains constant in the case of motion along different ellipses. And if we have several material bodies moving about S in ellipses with different semi-major axes $(a_1, a_2,...)$ and with different orbital periods $T_1, T_2 , ...,$ then

$$\frac{T^2_1}{a^3_1} = \frac{T^2_2}{a^3_2} = \cdots$$

whence follows Theorem IV.

Theorem V

If several material bodies are in uniform revolution about a centre of force, S, in circles with the centre at S and if the squares of their orbital periods are proportional to the cubes of the radii of their circular orbits, these material bodies are attracted to the centre, S, with forces that are inversely proportional to the square of the distances from S.

Proof

We utilize the expression for the force, F, derived in the proof of Theorem III and, noting that

$$a = b = P_0 S = r,$$

where r is the radius of the circle, we obtain

$$F = m \frac{4k^2}{r^3}.$$

This relationship connects the magnitude of the force with the area constant and the radius of the circle in which the body is moving. Since the area constant is related to the orbital period and the area of the circle by the equation $kT = \pi r^2$, the expression for F may be rewritten as:

$$F = m \frac{4\pi^2 r}{T^2}.^*$$

---

* It is possible to arrive at a similar result on the basis of the well-known equation for acceleration in uniform circular motion $w = v^2/r$. This equation was first derived by Huygens (1673) and, independently, by Newton.

If we have two bodies with masses m_1 and m_2 revolving in circles with radii r_1, r_2, the orbital periods of which are T_1 and T_2, then

$$\frac{F_1}{F_2} = \frac{m_1}{m_2} \cdot \frac{r_1 T_2^2}{r_2 T_1^2}.$$

And if Kepler's Third Law is satisfied:

$$\frac{T_2^2}{T_1^2} = \frac{r_2^3}{r_1^3}$$

then

$$\frac{F_1}{F_2} = \frac{m_1}{m_2} \cdot \frac{r_2^2}{r_1^2}.$$

Theorem VI

A particle placed without a material homogeneous spherical surface is attracted to its centre with a force inversely proportional to the square of the distance from its centre.

Proof

Fig. 57 (a and b) depicts two identical material spherical surfaces and the positions of material particles P and p attracted by these surfaces (in the figure, the spherical surfaces are shown only in part).

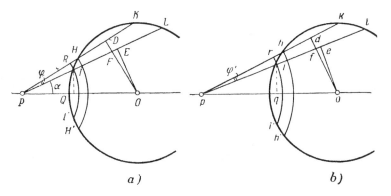

Fig. 57. Proof of Theorem VI

We draw the secants PHK, PIL and phk, pil sufficiently close to each other so that the corresponding chords on the two surfaces should be equal:

$$HK = hk, \quad IL = il.$$

We will then have the equalities: $DO = do$, $EO = eo$. From the similar triangles PIR and PFD we find

$$\frac{PI}{PF} = \frac{RI}{DF}$$

and from the similar triangles PIQ and PEO,

$$\frac{PI}{PO} = \frac{IQ}{EO}.$$

In the very same way we find analogous equalities for the other spherical surface (Fig. 57, b)

$$\frac{pf}{pi} = \frac{df}{ri} \quad \text{and} \quad \frac{po}{pi} = \frac{eo}{iq} = \frac{EO}{iq}.$$

Multiplying the left-hand and right-hand sides of the four equalities that have been derived, we obtain

$$\frac{PI^2 \cdot pf \cdot po}{pi^2 \cdot PF \cdot PO} = \frac{RI \cdot IQ \cdot df}{ri \cdot iq \cdot DF}.$$

It may be noted that in the limit, when angles φ and φ' between the secants tend to zero, the ratio df/DF tends to unity. In addition, it may be shown (this we leave to the reader) that by virtue of the equality $HK = hk$, the secants PH and ph will intersect the spherical surface at the same angles, that is, the angles between PH and ph and the tangents at points H and h are equal. The very same thing applies to the secants PI and pi. For this reason, if the arcs $\overset{\smile}{IH}$ and $\overset{\smile}{ih}$ are sufficiently small and if they may be identified with lengths of straight lines, the triangles RHI and rhi will be similar, that is, in the limit, when $H \to I$ and $h \to i$,

$$\frac{RI}{ri} = \frac{\overset{\smile}{HI}}{\overset{\smile}{hi}}.$$

The result is that in the limit

$$\frac{PI^2 \cdot pf \cdot po}{pi^2 \cdot PF \cdot PO} = \frac{\overset{\smile}{HI} \cdot IQ}{\overset{\smile}{hi} \cdot iq}. \tag{6}$$

Now let us examine the attraction of spherical shells of radii IQ and iq, which may be obtained by the rotation of arcs $\breve{H}I$ and $\breve{h}i$ about the diameters QB and qb.

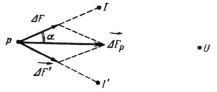

Fig. 58. Proof of Theorem VI

The forces with which individual elements of these shells attract points P and p are proportional to the masses of the shells (that is, to their areas) and inversely proportional to the square of the distance to them. Let us take symmetrically situated elementary sections on each of these shells, for example, near the points I, I', i, i'. The sections near I and I' attract the points P and p with forces ΔF and $\Delta F'$ (Fig. 58), equal in magnitude to $\mu \frac{\Delta \sigma}{PI^2}$ where μ is the constant of proportionality, and $\Delta \sigma$ is the area of each of these sections.

When $H \rightarrow I$ and $H' \rightarrow I'$, the directions of these forces in the limit coincide with the straight lines PI and $P'I'$. The total attraction of these two symmetrical elementary sections is directed along line PO to the centre of the sphere and is equal to ΔF_p. But $\Delta F_p = 2 \Delta F \cos \alpha$. Insofar as

$$\Delta F_P = \mu \frac{\Delta \sigma}{PI^2}, \quad \cos \alpha \frac{PQ}{PI},$$

we have

$$\Delta F_P = \mu \frac{2 \Delta \sigma}{PI^2} \frac{PQ}{PI}.$$

By summing, in pairs, the attraction of all elementary sections of this spherical shell we find that its attraction F_p is directed to the centre of the sphere O and equal to

$$F_P = \mu \frac{S}{PI^2} \frac{PQ}{PI}$$

where S is the area of this shell.

For a spherical shell in Fig. 57b we similarly find that its attraction is equal to

$$F_P = \mu \frac{S'}{pi^2} \frac{pq}{pi}.$$

Taking the ratio of these forces, we find that in the limit

$$\frac{F_P}{F_p}=\frac{S}{S'}\cdot\frac{pi^2}{PI^2}\cdot\frac{PQ}{pq}\cdot\frac{pi}{PI}\,.$$

Since the triangles PIQ and POF are similar in the limit,

$$\frac{PQ}{PI}=\frac{PF}{PO}.$$

Similarly,
$$\frac{pq}{pi}=\frac{pf}{po}.$$

The areas S and S' of our spherical shells are equal to $2\pi\ IQ\cdot\breve{HI}$ and $2\pi iq\cdot\breve{hi}$, respectively. And therefore

$$\frac{F_P}{F_p}=\frac{IQ\cdot\breve{HI}}{iq\cdot\breve{hi}}\cdot\frac{pi^2}{PI^2}\,\frac{PF\cdot po}{pf\cdot PO}\,.$$

Comparing this equality with equality (6) we find that

$$\frac{F_P}{F_p}=\frac{PI^2\cdot pf\cdot po}{pi^2\cdot PF\cdot PO}\cdot\frac{pi^2}{PI^2}\,\frac{PF\cdot po}{pf\cdot PO}=\frac{po^2}{PO^2}$$

which means that the forces of attraction of these spherical shells are inversely proportional to the squares of the distances of PO and po.

A similar result may be obtained for spherical shells constructed by the rotation of portions of arcs KL and kl.

Thus, the shells of spherical surfaces that have been isolated attract the material points P and p with forces that are inversely proportional to the distances of these points from the centres of the spherical surfaces. But we are able to divide the entire spherical surface into similar spherical shells. The attraction of each shell, and, hence, the attraction of the entire spherical surface is inversely proportional to the square of the distance of the attracting point from its centre.

Corollary from Theorem VI

A particle placed without a sphere with spherical distribution of densities is attracted to its centre with a force that is inversely proportional to the square of the distance to the centre of the sphere.

Proof

We divide the sphere into a large number of thin spherical layers which may be considered homogeneous. In accordance with Theorem VI, just proved above, each layer attracts a point without it with a force that is inversely proportional to the square of the distance of this point to the centre of the layer, which means that the attraction of the entire sphere, which is the integrated attraction of all the layers, will also be directed to the centre of the sphere and, in magnitude, it will be inversely proportional to the square of the distance from the centre of the sphere. Since the attraction of any material particles is proportional to their masses, it may be said that a material particle placed in the centre of a sphere and of mass equal to the mass of the sphere attracts exactly like the entire sphere.

A CATALOG OF SELECTED
DOVER BOOKS
IN SCIENCE AND MATHEMATICS

Physics

OPTICAL RESONANCE AND TWO-LEVEL ATOMS, L. Allen and J. H. Eberly. Clear, comprehensive introduction to basic principles behind all quantum optical resonance phenomena. 53 illustrations. Preface. Index. 256pp. 5⅜ x 8½. 0-486-65533-4

QUANTUM THEORY, David Bohm. This advanced undergraduate-level text presents the quantum theory in terms of qualitative and imaginative concepts, followed by specific applications worked out in mathematical detail. Preface. Index. 655pp. 5⅜ x 8½. 0-486-65969-0

ATOMIC PHYSICS (8th EDITION), Max Born. Nobel laureate's lucid treatment of kinetic theory of gases, elementary particles, nuclear atom, wave-corpuscles, atomic structure and spectral lines, much more. Over 40 appendices, bibliography. 495pp. 5⅜ x 8½. 0-486-65984-4

A SOPHISTICATE'S PRIMER OF RELATIVITY, P. W. Bridgman. Geared toward readers already acquainted with special relativity, this book transcends the view of theory as a working tool to answer natural questions: What is a frame of reference? What is a "law of nature"? What is the role of the "observer"? Extensive treatment, written in terms accessible to those without a scientific background. 1983 ed. xlviii+172pp. 5⅜ x 8½. 0-486-42549-5

AN INTRODUCTION TO HAMILTONIAN OPTICS, H. A. Buchdahl. Detailed account of the Hamiltonian treatment of aberration theory in geometrical optics. Many classes of optical systems defined in terms of the symmetries they possess. Problems with detailed solutions. 1970 edition. xv + 360pp. 5⅜ x 8½. 0-486-67597-1

PRIMER OF QUANTUM MECHANICS, Marvin Chester. Introductory text examines the classical quantum bead on a track: its state and representations; operator eigenvalues; harmonic oscillator and bound bead in a symmetric force field; and bead in a spherical shell. Other topics include spin, matrices, and the structure of quantum mechanics; the simplest atom; indistinguishable particles; and stationary-state perturbation theory. 1992 ed. xiv+314pp. 6⅛ x 9¼. 0-486-42878-8

LECTURES ON QUANTUM MECHANICS, Paul A. M. Dirac. Four concise, brilliant lectures on mathematical methods in quantum mechanics from Nobel Prize-winning quantum pioneer build on idea of visualizing quantum theory through the use of classical mechanics. 96pp. 5⅜ x 8½. 0-486-41713-1

THIRTY YEARS THAT SHOOK PHYSICS: THE STORY OF QUANTUM THEORY, George Gamow. Lucid, accessible introduction to influential theory of energy and matter. Careful explanations of Dirac's anti-particles, Bohr's model of the atom, much more. 12 plates. Numerous drawings. 240pp. 5⅜ x 8½. 0-486-24895-X

ELECTRONIC STRUCTURE AND THE PROPERTIES OF SOLIDS: THE PHYSICS OF THE CHEMICAL BOND, Walter A. Harrison. Innovative text offers basic understanding of the electronic structure of covalent and ionic solids, simple metals, transition metals and their compounds. Problems. 1980 edition. 582pp. 6⅛ x 9¼. 0-486-66021-4

CATALOG OF DOVER BOOKS

HYDRODYNAMIC AND HYDROMAGNETIC STABILITY, S. Chandrasekhar. Lucid examination of the Rayleigh-Benard problem; clear coverage of the theory of instabilities causing convection. 704pp. 5⅜ x 8¼. 0-486-64071-X

INVESTIGATIONS ON THE THEORY OF THE BROWNIAN MOVEMENT, Albert Einstein. Five papers (1905–8) investigating dynamics of Brownian motion and evolving elementary theory. Notes by R. Fürth. 122pp. 5⅜ x 8½. 0-486-60304-0

THE PHYSICS OF WAVES, William C. Elmore and Mark A. Heald. Unique overview of classical wave theory. Acoustics, optics, electromagnetic radiation, more. Ideal as classroom text or for self-study. Problems. 477pp. 5⅜ x 8½. 0-486-64926-1

GRAVITY, George Gamow. Distinguished physicist and teacher takes reader-friendly look at three scientists whose work unlocked many of the mysteries behind the laws of physics: Galileo, Newton, and Einstein. Most of the book focuses on Newton's ideas, with a concluding chapter on post-Einsteinian speculations concerning the relationship between gravity and other physical phenomena. 160pp. 5⅜ x 8½. 0-486-42563-0

PHYSICAL PRINCIPLES OF THE QUANTUM THEORY, Werner Heisenberg. Nobel Laureate discusses quantum theory, uncertainty, wave mechanics, work of Dirac, Schroedinger, Compton, Wilson, Einstein, etc. 184pp. 5⅜ x 8½. 0-486-60113-7

ATOMIC SPECTRA AND ATOMIC STRUCTURE, Gerhard Herzberg. One of best introductions; especially for specialist in other fields. Treatment is physical rather than mathematical. 80 illustrations. 257pp. 5⅜ x 8½. 0-486-60115-3

AN INTRODUCTION TO STATISTICAL THERMODYNAMICS, Terrell L. Hill. Excellent basic text offers wide-ranging coverage of quantum statistical mechanics, systems of interacting molecules, quantum statistics, more. 523pp. 5⅜ x 8½. 0-486-65242-4

THEORETICAL PHYSICS, Georg Joos, with Ira M. Freeman. Classic overview covers essential math, mechanics, electromagnetic theory, thermodynamics, quantum mechanics, nuclear physics, other topics. First paperback edition. xxiii + 885pp. 5⅜ x 8½. 0-486-65227-0

PROBLEMS AND SOLUTIONS IN QUANTUM CHEMISTRY AND PHYSICS, Charles S. Johnson, Jr. and Lee G. Pedersen. Unusually varied problems, detailed solutions in coverage of quantum mechanics, wave mechanics, angular momentum, molecular spectroscopy, more. 280 problems plus 139 supplementary exercises. 430pp. 6½ x 9¼. 0-486-65236-X

THEORETICAL SOLID STATE PHYSICS, Vol. 1: Perfect Lattices in Equilibrium; Vol. II: Non-Equilibrium and Disorder, William Jones and Norman H. March. Monumental reference work covers fundamental theory of equilibrium properties of perfect crystalline solids, non-equilibrium properties, defects and disordered systems. Appendices. Problems. Preface. Diagrams. Index. Bibliography. Total of 1,301pp. 5⅜ x 8½. Two volumes. Vol. I: 0-486-65015-4 Vol. II: 0-486-65016-2

WHAT IS RELATIVITY? L. D. Landau and G. B. Rumer. Written by a Nobel Prize physicist and his distinguished colleague, this compelling book explains the special theory of relativity to readers with no scientific background, using such familiar objects as trains, rulers, and clocks. 1960 ed. vi+72pp. 5⅜ x 8½. 0-486-42806-0

A TREATISE ON ELECTRICITY AND MAGNETISM, James Clerk Maxwell. Important foundation work of modern physics. Brings to final form Maxwell's theory of electromagnetism and rigorously derives his general equations of field theory. 1,084pp. 5⅜ x 8½. Two-vol. set. Vol. I: 0-486-60636-8 Vol. II: 0-486-60637-6

QUANTUM MECHANICS: PRINCIPLES AND FORMALISM, Roy McWeeny. Graduate student-oriented volume develops subject as fundamental discipline, opening with review of origins of Schrödinger's equations and vector spaces. Focusing on main principles of quantum mechanics and their immediate consequences, it concludes with final generalizations covering alternative "languages" or representations. 1972 ed. 15 figures. xi+155pp. 5⅜ x 8½. 0-486-42829-X

INTRODUCTION TO QUANTUM MECHANICS With Applications to Chemistry, Linus Pauling & E. Bright Wilson, Jr. Classic undergraduate text by Nobel Prize winner applies quantum mechanics to chemical and physical problems. Numerous tables and figures enhance the text. Chapter bibliographies. Appendices. Index. 468pp. 5⅜ x 8½. 0-486-64871-0

METHODS OF THERMODYNAMICS, Howard Reiss. Outstanding text focuses on physical technique of thermodynamics, typical problem areas of understanding, and significance and use of thermodynamic potential. 1965 edition. 238pp. 5⅜ x 8½. 0-486-69445-3

THE ELECTROMAGNETIC FIELD, Albert Shadowitz. Comprehensive undergraduate text covers basics of electric and magnetic fields, builds up to electromagnetic theory. Also related topics, including relativity. Over 900 problems. 768pp. 5⅝ x 8¼. 0-486-65660-8

GREAT EXPERIMENTS IN PHYSICS: FIRSTHAND ACCOUNTS FROM GALILEO TO EINSTEIN, Morris H. Shamos (ed.). 25 crucial discoveries: Newton's laws of motion, Chadwick's study of the neutron, Hertz on electromagnetic waves, more. Original accounts clearly annotated. 370pp. 5⅜ x 8½. 0-486-25346-5

EINSTEIN'S LEGACY, Julian Schwinger. A Nobel Laureate relates fascinating story of Einstein and development of relativity theory in well-illustrated, nontechnical volume. Subjects include meaning of time, paradoxes of space travel, gravity and its effect on light, non-Euclidean geometry and curving of space-time, impact of radio astronomy and space-age discoveries, and more. 189 b/w illustrations. xiv+250pp. 8⅜ x 9¼. 0-486-41974-6

STATISTICAL PHYSICS, Gregory H. Wannier. Classic text combines thermodynamics, statistical mechanics and kinetic theory in one unified presentation of thermal physics. Problems with solutions. Bibliography. 532pp. 5⅜ x 8½. 0-486-65401-X

Paperbound unless otherwise indicated. Available at your book dealer, online at **www.doverpublications.com**, or by writing to Dept. GI, Dover Publications, Inc., 31 East 2nd Street, Mineola, NY 11501. For current price information or for free catalogues (please indicate field of interest), write to Dover Publications or log on to **www.doverpublications.com** and see every Dover book in print. Dover publishes more than 500 books each year on science, elementary and advanced mathematics, biology, music, art, literary history, social sciences, and other areas.